THE PHILOSOPHY OF SPACE & TIME

BY

HANS REICHENBACH

Translated by

MARIA REICHENBACH

AND

JOHN FREUND

With Introductory Remarks by

RUDOLF CARNAP

DOVER PUBLICATIONS, INC.
NEW YORK

This Dover edition, first published in 1957, is a new English translation of *Philosophie der Raum-Zeit-Lehre.*

Library of Congress Catalog Card Number: 58-7082
International Standard Book Number

ISBN-13: 978-0-486-60443-5
ISBN-10: 0-486-60443-8

Manufactured in the United States by Courier Corporation
60443825 2014
www.doverpublications.com

ACKNOWLEDGEMENT

The English translation of this book was partly checked by my late husband. In the preparation of the manuscript for publication valuable help was given me by Professor Rudolf Carnap, Dr. Wesley C. Salmon, and Dr. Bruce Taylor.

My special thanks go to Mrs. Ruth Anna Mathers, who for many weeks worked patiently with me on the final version.

MARIA REICHENBACH

Los Angeles
July 1956

PREFACE

The publication of this book gives me the welcome opportunity to express my appreciation to the *Notgemeinschaft der deutschen Wissenschaft* which has enabled me for a number of years to carry on my work in the philosophical analysis of physics.

I thank Dr. Rudolf Carnap, Vienna, and Dr. Kurt Grelling, Berlin, for their friendly criticisms concerning some details, and Mr. Hans Stotz and Mr. Martin Strauss for their valuable help in the proofreading of the manuscript.

HANS REICHENBACH

Berlin,
October 1927

INTRODUCTORY REMARKS TO THE ENGLISH EDITION

Since ancient times the question of the nature of geometry has been a decisive problem for any theory of knowledge. The principles of geometry, e.g., Euclid's axioms, seem to possess two characteristics which are not easily reconciled. On the one hand, they appear as immediately evident and therefore to hold with necessity. On the other hand, their validity is not purely logical but factual; in technical terms, they are not analytic but synthetic. This is shown by the fact that, on the basis of certain measurements of angles and lengths of physical bodies the results of other measurements can be predicted. Kant boldly accepted the conjunction of both characteristics: from the apparently necessary validity of the principles of geometry he concluded that their knowledge is *a priori* (i.e., independent of experience) although they are synthetic. When mathematicians constructed about a hundred years ago systems of non-Euclidean geometries, a controversy arose about the method of determining which of the systems, one Euclidean and infinitely many non-Euclidean, holds for the space of physics. Gauss was the first to suggest that the determination should be made by physical measurements. But the great majority of philosophers throughout the last century maintained the Kantian doctrine that geometry is independent of experience.

At the beginning of our century Poincaré pointed out the following new aspect of the situation. No matter what observational facts are found, the physicist is free to ascribe to physical space any one of the mathematically possible geometrical structures, provided he makes suitable adjustments in the laws of mechanics and optics and consequently in the rules for measuring length. This was an important insight. But Poincaré went further and asserted that physicists would always choose the Euclidean structure because of its simplicity. History refuted this prediction only a few years later, when Einstein used a certain non-Euclidean geometry in his general theory of relativity. Hereby he obtained a considerable gain in simplicity for the total system of physics in spite of the loss in simplicity for geometry.

Through this development it has become clear that the situation concerning the nature of geometry is as follows. It is necessary to distinguish between pure or mathematical geometry and physical geometry. The statements of pure geometry hold logically, but they deal only with abstract structures and say nothing about physical space. Physical geometry describes the structure of physical space; it is a part of physics. The validity of its statements is to be established empirically—as it has to be in any other part of physics—after rules for measuring the magnitudes involved, especially length, have been stated. (In Kantian terminology, mathematical geometry holds indeed *a priori*, as Kant asserted, but only because it is analytic. Physical geometry is indeed synthetic; but it is based on experience and hence does not hold *a priori*. In neither of the two branches of science which are called "geometry" do synthetic judgments *a priori* occur. Thus Kant's doctrine must be abandoned.)

In physical geometry, there are two possible procedures for establishing a theory of physical space. First, the physicist may freely choose the rules for measuring length. After this choice is made, the question of the geometrical structure of physical space becomes empirical; it is to be answered on the basis of the results of experiments. Alternatively, the physicist may freely choose the structure of physical space; but then he must adjust the rules of measurement in view of the observational facts. (Although Poincaré emphasized the second way, he also saw the first clearly. This point seems to be overlooked by those philosophers, among them Reichenbach, who regard Poincaré's view on geometry as non-empiricist and purely conventionalist.)

The view just outlined concerning the nature of geometry in physics stresses, on the one hand, the empirical character of physical geometry and, on the other hand, recognizes the important function of conventions. This view was developed in the twenties of our century by those philosophers who studied the logical and methodological problems connected with the theory of relativity, among them Schlick, Reichenbach, and myself. The first comprehensive and systematic representation of this conception was given by Reichenbach in 1928 in his *Philosophie der Raum-Zeit-Lehre* (the original of the present translation). This work was an important landmark in the development of the empiricist conception of geometry. In my judgment it is still the best book in the field. Therefore the appearance of an English edition is to be highly welcomed; it satisfies a definite need, all the more since the German original is out of print.

The book deals with the problems of the foundations of geometry—and also of the theory of time, closely connected with that of space by Einstein's conception—in all their various aspects, e.g., the relations between theory and observations, connected by coordinative definitions, the relations between topological and metrical properties of space, and also the psychological problem of the possibility of a visual intuition of non-Euclidean structures.

Of the many fruitful ideas which Reichenbach contributed to the development of this philosophical theory, I will mention only one, which seems to me of great interest for the methodology of physics but which has so far not found the attention it deserves. This is the principle of the elimination of universal forces. Reichenbach calls those physical forces universal which affect all substances in the same way and against which no isolating walls can be built. Let T be the form of Einstein's theory which uses that particular non-Euclidean structure of space which Einstein proposes; in T there are no universal forces. According to our above discussion, T can be transformed into another form T' which is physically equivalent with T in the sense of yielding the same observable results, but uses a different geometrical structure. Reichenbach shows that any such theory T' has to assume that our measuring rods undergo contractions or expansions depending merely upon their positions in space, and hence has to introduce universal forces to account for these changes. Reichenbach proposes to accept as a general methodological principle that we choose that form of a theory among physically equivalent forms (or, in other words, that definition of "rigid body" or "measuring standard") with respect to which all universal forces disappear. If this principle is accepted, the arbitrariness in the choice of a measuring procedure is avoided and the question of the geometrical structure of physical space has a unique answer, to be determined by physical measurements.

Even more outstanding than the contributions of detail in this book is the spirit in which it was written. The constant careful attention to scientifically established facts and to the content of the scientific hypotheses to be analyzed and logically reconstructed, the exact formulation of the philosophical results, and the clear and cogent presentation of the arguments supporting them, make this work a model of scientific thinking in philosophy.

RUDOLF CARNAP

University of California
at Los Angeles
July, 1956

TABLE OF CONTENTS

Chapter III
Space and Time

A. The Space-Time Manifold without Gravitational Fields

B. Gravitation Filled Space-Time Manifolds

C. The Most General Properties of Space and Time

INTRODUCTION

If the philosophical method of our time is compared with the method of the great system builders of the 17th and 18th centuries, a fundamental difference in the respective attitudes to the natural sciences becomes evident. The classical philosophers had a close connection with the science of their times; some of them, such as Descartes and Leibniz, were leading mathematicians and physicists themselves. More recently philosophy and science have become estranged, a situation which has led to an unproductive tension between the two groups. The philosophers, whose professional training has usually been acquired in the pursuit of historical and philological studies, accuse the scientist of too much specialization and turn instead to metaphysical problems; the scientists, on the other hand, miss in philosophy the treatment of epistemological problems, which, though solved by a Leibniz or Kant within the framework of the science of their time, demand a fresh analysis within the framework of contemporary science. This alienation is expressed in mutual contempt in which each misunderstands the purposes of the other's endeavors.

Looking back into history one can trace the roots of this division throughout the past century. For Kant, knowledge as realized in mathematical physics was still the starting point of epistemology; although this basis constitutes a certain one-sidedness of his system, it also accounts for the strength of his epistemological position to which his philosophy owes its great influence. It is surprising, however, how little use Kant made of particular scientific results in the elaboration of his system, how little scientific material he employed in his main epistemological works, even in the form of examples. He must have seen the scientific conception of knowledge as a whole and created his system out of this experience, which produced, as the result of an analysis of pure reason, the very conception of knowledge of the mathematical physics of his time. How well he must have explicated this conception of knowledge can be inferred from the vivid interest expressed by the natural scientists. Whether they were adversaries or adherents, a clarification of their position with respect to that of Kant seemed to be natural and necessary to them, and they gradually identified Kant's doctrine with philosophy as such. But Kant's solution of the epistemological problem was at the same time the last one in which science played a role. Later philosophical systems had no longer any connection with the science of their time; and though some of them, such as Schelling's and Hegel's natural philosophy, treated scientific material to a larger extent than Kant, their philosophy of nature is a naive evaluation of scientific results rather than a true understanding of the spirit of scientific research. Since then science and philosophy have remained separated. The speculative and the rationalist-analytic components of Kant's system were preserved, while the relation to science was renounced. The philosopher allied himself with the humanities; so far as science interested him at all, he surmised that the problem of science had been solved since Kant and that a further development of science consisted only in filling out Kant's program, a conception which even in the more flexible form of the Neo-Kantian school could not be prevented from coming into conflict with the actual development of science. Science, in the meantime, went its own way. Certainly, one cannot reproach Kant for not anticipating this development, but neither can one expect the modern scientist to acknowledge Kant's philosophy as the basis of his own epistemology. Neither in Kant nor in the prevailing schools of philosophy does he find an epistemology that enables him to understand his own scientific activity. Philosophy still acts like a stranger toward the gigantic complex of natural science, even to the point of rejecting it.

In the course of the last century the scientists themselves elaborated the epistemological foundations as well as the content of scientific theories. Of course, only a few outstanding men were conscious of the philosophical character of their methodology; most of the results were achieved inadvertently without intending philosophical solutions, in the pursuit of specific scientific interests which, however, were bound to lead to philosophical inquiries. Thus we are faced with the strange result that during the last century an exact theory of knowledge was constructed, not by philosophers, but by scientists, and that in the pursuit of particular scientific investigations more epistemology was produced than in the process of philosophical speculation. And the problems thus solved were truly epistemological problems. If the speculatively oriented philosophy of our time denies to contemporary science its philosophical character, if it calls contributions such as the theory of relativity or the theory of sets unphilosophical and belonging in the special sciences, this judgment expresses only the inability to perceive the philosophical content of modern scientific thought. Today mathematical physics, by means of its infinitely refined mathematical and experimental methods, treats the same problems that constituted the foundation of the epistemology of Descartes, Leibniz and Kant. Yet an adequate insight into the techniques of scientific inquiry is necessary in order to understand what a powerful instrument for the analysis of basic philosophical questions has been created and what potentialities for philosophical exploitation it contains.

Gradually, however, the situation has become too complicated for the scientist. He can no longer work out the actual philosophical implications, for the simple reason that one individual is not capable of carrying on scientific and philosophical work at the same time. A division of labor seems inevitable, since empirical as well as epistemological research demands an amount of detailed work that surpasses the capacity of one individual. It should also be mentioned that the philosophical and scientific goals, though in general depending upon each other, oppose each other within the mentality of the individual scholar. The philosophic analysis of the meaning and significance of scientific statements can almost hinder the processes of scientific research and paralyze the pioneering spirit, which would lack the courage to walk new paths without a certain amount of irresponsibility. The style of modern science has gradually adopted the hurried pace of technology induced by competition; one might regret this mechanistic trend, but it seems to be the necessary form of modern productivity. We cannot counteract this tendency by a competition with less technical means, but solely by means of a philosophical analysis of the process of knowledge itself; it is the discovery of the significance of this machine age knowledge which, in the minds of many people, will remain mere technology, but which, in its system as a whole, reveals a depth of insight that can only be reached through the teamwork of an organized group of individual scholars.

To carry out such a philosophy of nature must therefore remain the prerogative of a special group of individual scholars such as has recently emerged, of a group that on the one hand masters the technique of mathematical science, and on the other hand is not weighed down by it to such an extent as to lose its philosophic perspective over details. For in the same way as philosophic contemplation can inhibit the daring step of the scientific investigator, specialized research can limit the ability for philosophic interpretation. The reproach by philosophers that scientists lack an understanding of philosophical problems is no less justified than the one voiced by the other side charging a lack of understanding of scientific problems.

From this circumstance one should not draw the conclusion, however, that one ought to carry on philosophy in a speculative vein, apart from the sciences. On the contrary, one should approach science from a philosophical point of view and try to construct with its sharpened tools the philosophy of this technically refined knowledge.

From this point of view, the author has carried through a number of investigations exploring the complex of mathematical physics from various directions. The natural organization of this basic science led to the decision to present the investigations concerning the problems of time and space as a separate unit; an exposition of further studies will follow. For the theory of space and time comprehensive material was available, arising on the one hand from the mathematical analysis of geometry, on the other hand from Einstein's theory of relativity. This theory provides a vivid example of the fruitfulness of physical questions for philosophical explication. Thus a philosophy of space and time is nowadays always a philosophy of relativity—this duality probably characterizes best the method of scientific analysis which is the basis of such a philosophy.

It seemed necessary to include an exposition of the material in our presentation. A mere reference to mathematico-physical publications of the material would be inadequate, because all these books are geared too much to a mathematico-physical interpretation and neglect the philosophical foundations. On the other hand, it seemed to be out of the question to enter into a philosophical evaluation of this material without keeping it clearly in mind at every moment. Modern philosophy of nature will have to develop in as close a connection with the actual natural and mathematical sciences as has been taken for granted for cultural philosophy and its historical subject matter. And if we permit historians of philosophy to quote repeatedly, in their presentations, parts of the original historical text whose content cannot be completely exhausted by being paraphrased or translated, we should not be surprised that in his philosophical investigations of nature the philosopher will go back to the original mathematical language in which the "book of nature" is written, since mathematical language can even less be exhausted by being paraphrased or translated. A considerable part of the necessary mathematical work was completed in the author's *Axiomatik der relativistischen Raum-Zeit-Lehre*[1] and detailed mathematical computations could therefore be omitted from this book. The philosophical interpretation of the theory of space and time presupposes the earlier work to which I have to refer the reader for rigorous proofs of many statements in the present book. But the occasional use of a mathematical formula occurring in the text will throw even more light upon the epistemological foundation.

The path of the present philosophical work led therefore through the natural sciences, yet the wealth of mathematical and physical material did not appear as an obstacle, but rather as an inexhaustible source of further philosophical insights. It is hoped, in this manner, to give an example of the superiority of a philosophical method closely connected with the results of empirical science. All the detailed mathematical work achieved by outstanding men is at its disposal and becomes systematically integrated from the viewpoint of the philosopher. Formulations whose universality would not mean anything in isolation acquire the utmost significance, if they are supported by a detailed analysis of particular instances which they generalize. Modern scientific epistemology therefore justifies discoveries of such far-reaching consequences as would, in former times, have been merely empty speculation, phantasies without empirical foundation. It is characteristic of this emerging scientific trend in philosophy to emphasize the combination of detailed work with an overall comprehensiveness of the problem; whoever charges it with narrow-mindedness or sterility shows only that he confuses rigor of method with narrowness of aim.

This book has been written in the knowledge that solutions are attainable. It is intended, at the same time, to present in a comprehensive fashion the treasure of philosophical results that has become the common property of scientific philosophy, constituting already a certain common tradition, and also to go beyond it on new paths that were opened to the author through a persistent analysis of mathematical physics. If in this survey, therefore, ideas are not always traced back to their authors, this will best be understood by a person who is a collaborator in this field and who knows how frequently ideas are oscillating today between various thinkers until they find their ultimate

[1] Braunschweig 1924, Friedrich Vieweg and Son, A.G. (in the following referred to as A.).

formulation as a product of teamwork. This accumulation of common knowledge is the characteristic mark of the new philosophical orientation, which, due to its origin in the empirical sciences, stands even methodologically in contrast to the isolated systems of the speculative philosophers and gains its superiority from this source. Philosophy of science is not intended to be one of those systems that originate in the mind of a lonely thinker and stand like marble monuments before the gaze of generations, but should be considered a science like the other sciences, a fund of cooperatively discovered propositions whose acceptance, independent of the framework of a system, can be required from anybody interested in these matters. The meaning of concepts may vary, of course, depending upon the context in which they are used; but this kind of ambiguity can be avoided by making language more precise and need not lead to a renunciation of objective philosophical knowledge altogether. If the effect of the philosophy of systems was to destroy the concept of philosophical truth and replace it by the concept of consistency within the system, one may see as the noblest aim of scientific philosophy the establishment of the concept of objective truth as the ultimate criterion of all philosophical knowledge.

THE PHILOSOPHY OF

SPACE & TIME

CHAPTER I. SPACE

§ 1. THE AXIOM OF THE PARALLELS AND NON-EUCLIDEAN GEOMETRY

In Euclid's work, the geometrical achievements of the ancients reached their final form: geometry was established as a closed and complete system. The basis of the system was given by the geometrical axioms[1], from which all theorems were derived. The great practical significance of this construction consisted in the fact that it endowed geometry with a certainty never previously attained by any other science. The small number of axioms forming the foundation of the system were so self-evident that their truth was accepted without reservation. The entire construction of geometry was carried through by a skillful combination of the axioms alone, without any addition of further assumptions; the reliability of the logical inferences used in the proofs was so great that the derived theorems, which were sometimes quite involved, could be regarded as certain as the axioms. Geometry thus became the prototype of a demonstrable science, the first instance of a scientific rigor which, since that time, has been the ideal of every science. In particular, the philosophers of all ages have regarded it as their highest aim to prove their conclusions "by the geometrical method."

Euclid's axiomatic construction was also important in another respect. The problem of demonstrability of a science was solved by Euclid in so far as he had reduced the science to a system of axioms. But now arose the epistemological question how to justify the truth of those first assumptions. If the certainty of the axioms was transferred

[1] Euclid distinguished between axioms, postulates and definitions. We may be allowed for our present purpose to include all these concepts under the name of axioms.

to the derived theorems by means of the system of logical concatenations, the problem of the truth of this involved construction was transferred, conversely, to the axioms. It is precisely the assertion of the truth of the axioms which epitomizes the problem of scientific knowledge, once the connection between axioms and theorems has been carried through. In other words: the *implicational* character of mathematical demonstrability was recognized, i.e., the undeniable fact that only the implication "if *a*, then *b*" is accessible to logical proof. The problem of the categorical assertion "*a* is true *b* is true", which is no longer tied to the "if", calls for an independent solution. The truth of the axioms, in fact, represents the intrinsic problem of every science. The axiomatic method has not been able to establish knowledge with absolute certainty; it could only reduce the question of such knowledge to a precise thesis and thus present it for philosophical discussion.

This effect of the axiomatic construction, however, was not recognized until long after Euclid's time. Precise epistemological formulations could not be expected from a naive epoch, in which philosophy was not yet based upon well-developed special sciences, and thinkers concerned themselves with cruder things than the truth of simple and apparently self-evident axioms. Unless one was a skeptic, one was content with the fact that certain assumptions had to be believed axiomatically; analytical philosophy has learned mainly through Kant's critical philosophy to discover genuine problems in questions previously utilized only by skeptics in order to deny the possibility of knowledge. These questions became the central problems of epistemology. For two thousand years the criticism of the axiomatic construction has remained within the frame of mathematical questions, the elaboration of which, however, led to peculiar discoveries, and eventually called for a return to philosophical investigations.

The mathematical question concerned the reducibility of the axiomatic system, i.e., the problem whether Euclid's axioms represented ultimate propositions or whether there was a possibility of reducing them to still simpler and more self-evident statements. Since the individual axioms were quite different in character with respect to their immediacy, the question arose whether some of the more complicated axioms might be conceived as consequences of the simpler ones, i.e., whether they could be included among the theorems. In particular, the demonstrability of the axiom of the parallels was investigated. This axiom states that through a given point there is *one and only one* parallel to a given straight line (which does not go through the given point), i.e., one straight line which lies in the same plane with the first one and does not intersect it. At first glance this axiom appears to be self-evident. There is, however, something unsatisfactory about it, because it contains a statement about infinity; the assertion that the two lines do not intersect within a finite distance transcends all possible experience. The demonstrability of this axiom would have enhanced the certainty of geometry to a great extent, and the history of mathematics tells us that excellent mathematicians from Proclus to Gauss have tried in vain to solve the problem.

A new turn was given to the question through the discovery that it was possible to do without the axiom of parallels altogether. Instead of proving its truth the opposite method was employed: it was demonstrated that this axiom could be dispensed with. Although the existence of several parallels to a given line through one point contradicts the human power of visualization, this assumption could be introduced as an axiom, and a consistent geometry could be developed in combination with Euclid's other axioms. This discovery was made almost simultaneously in the twenties of the last century by the Hungarian, Bolyai, and the Russian, Lobatschewsky; Gauss is said to have conceived the idea somewhat earlier without publishing it.

But what can we make of a geometry that assumes the opposite of the axiom of the parallels? In order to understand the possibility of a non-Euclidean geometry, it must be remembered that the axiomatic construction furnishes the proof of a statement in terms of logical derivations from the axioms alone. The drawing of a figure is only a means to assist visualization, but is never used as a factor in the proof; we know that a proof is also possible by the help of "badly-drawn" figures in which so-called congruent triangles have sides obviously different in length. It is not the immediate picture of the figure, but a concatenation of logical relations that compels us to accept the proof. This consideration holds equally well for non-Euclidean geometry; although the drawing looks like a "badly-drawn" figure, we can with its help discover whether the logical requirements have been satisfied, just as we can do in Euclidean geometry. This is why non-Euclidean geometry has been developed from its inception in an axiomatic construction; in contradistinction to Euclidean geometry where the theorems were known first and the axiomatic foundation was developed later, the axiomatic construction was the instrument of discovery in non-Euclidean geometry.

With this consideration, which was meant only to make non-Euclidean geometry plausible, we touch upon the problem of the *visualization of geometry.* Since this question will be treated at greater length in a later section, the remark about "badly-drawn" figures should be taken as a passing comment. What was intended was to stress the fact that the essence of a geometrical proof is contained in the logic of its derivations, not in the proportions of the figures. Non-Euclidean geometry is a logically constructible system—this was the first and most important result established by its inventors.

It is true that a strict proof was still missing. No contradictions were encountered—yet did this mean that none would be encountered in the future? This question constitutes the fundamental problem concerning an axiomatically constructed logical system. It is to be expected that non-Euclidean statements directly contradict those of Euclidean geometry; one must not be surprised if, for instance, the sum of the angles of a triangle is found to be smaller than two right angles. This contradiction follows necessarily from the reformulation of the axiom of the parallels. What is to be required is that the new geometrical system be self-consistent. The possibility can be imagined that a statement *a*, proved within the non-Euclidean axiomatic system, is not tenable in a later development, i.e., that the statement *not-a* as well as the statement *a* is provable in the axiomatic system. It was incumbent upon the early adherents of non-Euclidean geometry, therefore, to prove that such a contradiction could never happen.

The proof was furnished to a certain extent by Klein's[1] Euclidean model of non-Euclidean geometry. Klein succeeded in coordinating the concepts of Euclidean geometry, its points, straight lines, and planes, its concept of congruence, etc., to the corresponding concepts of non-Euclidean geometry, so that every statement of one geometry corresponds to a statement of the other. If in non-Euclidean geometry a statement *a* and also a statement *not-a* could be proved, the same would hold for the coordinated statements *a′* and *not-a′* of Euclidean geometry; a contradiction in non-Euclidean geometry would entail a corresponding contradiction in Euclidean geometry. The result was a proof of consistency, the first in the history of mathematics: it proceeds by reducing a new system of statements to an earlier one, the consistency of which is regarded as virtually certain.[2]

After these investigations by Klein the mathematical significance of

[1] For a more detailed presentation see § 11.

[2] Hilbert later proved the consistency of Euclidean geometry by a reduction to arithmetic. The consistency of arithmetic, which can no longer be proved by reduction, needs a separate proof; this most important problem, which has found an elaborate treatment by Hilbert and his school, is still under discussion.

non-Euclidean geometry was recognized.[1] Compared with the natural geometry of Euclid, that of Bolyai and Lobatschewsky appeared strange and artificial; but its mathematical legitimacy was beyond question. It turned out later that another kind of non-Euclidean geometry was possible. The axiom of the parallels in Euclidean geometry asserts that to a given straight line through a given point there exists exactly one parallel; apart from the device used by Bolyai and Lobatschewsky to deny this axiom by assuming the existence of several parallels, there was a third possibility, that of denying the existence of any parallel. However, in order to carry through this assumption consistently,[2] a certain change in a number of Euclid's other axioms referring to the infinity of a straight line was required. By the help of these changes it became possible to carry through this new type of non-Euclidean geometry.

As a result of these developments there exists not one geometry but a plurality of geometries. With this mathematical discovery, the epistemological problem of the axioms was given a new solution. If mathematics is not required to use certain systems of axioms, but is in a position to employ the axiom *not-a* as well as the axiom *a*, then the assertion *a* does not belong in mathematics, and mathematics is solely the science of implication, i.e., of relations of the form "if . . . then"; consequently, for geometry as a mathematical science, there is no problem concerning the truth of the axioms. This apparently unsolvable problem turns out to be a pseudo-problem. The axioms are not true or false, but arbitrary statements. It was soon discovered that the other axioms could be treated in the same way as the axiom of the parallels. "Non-Archimedian," "non-Pascalian," etc., geometries were constructed; a more detailed exposition will be found in § 14.

These considerations leave us with the problem into which discipline the question of the truth of the assertion *a* should be incorporated.

[1] Klein did not start his investigations with the avowed purpose of establishing a proof of consistency; the proof came about inadvertently, so to speak, as a result of the construction of the model carried out with purely mathematical intentions. L. Bieberbach has shown recently that the recognition of the significance of non-Euclidean geometry was the result of long years of struggle. *Berl. Akademieber.* 1925, phys.-math. Klasse, p. 381. See Bonola-Liebmann, *Nichteuklidische Geometrie*, Leipzig 1921 and Engel-Stäckel, *Theorie der Parallellinien von Euklid bis Gauss*, Leipzig 1895, for the earlier history of the axiom of the parallels.

[2] The axiom of the parallels is independent of the other axioms of Euclid only in so far as it asserts the existence of at most one parallel; that there exists at least one parallel can be demonstrated in terms of the other axioms. This fact is stated with masterful precision in Euclid's work.

Nobody can deny that we regard this statement as meaningful; common sense is convinced that real space, the space in which we live and move around, corresponds to the axioms of Euclid and that with respect to this space *a* is true, while *not-a* is false. The discussion of this statement leads away from mathematics; as a question about a property of the physical world, it is a *physical* question, not a *mathematical* one. This distinction, which grew out of the discovery of non-Euclidean geometry, has a fundamental significance: it divides the problem of space into two parts; the problem of mathematical space is recognized as different from the problem of physical space.

It will be readily understood that the philosophical insight into the twofold nature of space became possible only after mathematics had made the step from Euclid's geometry to non-Euclidean geometries. Up to that time physics had assumed the axioms of geometry as the self-evident basis of its description of nature. If several kinds of geometries were regarded as mathematically equivalent, the question arose which of these geometries was applicable to physical reality; there is no necessity to single out Euclidean geometry for this purpose. Mathematics shows a variety of possible forms of relations among which physics selects the real one by means of observations and experiments. Mathematics, for instance, teaches how the planets would move if the force of attraction of the sun should decrease with the second or third or *n*th power of the distance; physics decides that the second power holds in the real world. With respect to geometry there had been a difference; only *one* kind of geometry had been developed and the problem of choice among geometries had not existed. After the discoveries of non-Euclidean geometries the duality of *physical* and *possible* space was recognized. Mathematics reveals the possible spaces; physics decides which among them corresponds to physical space. In contrast to all earlier conceptions, in particular to the philosophy of Kant, it becomes now a task of physics to determine the geometry of physical space, just as physics determines the shape of the earth or the motions of the planets, by means of observations and experiments.

But what methods should physics employ in order to come to a decision? The answer to this question will at the same time supply an answer to the question why we are justified in speaking of a specific physical space. Before this problem can be investigated more closely, another aspect of geometry will have to be discussed. For physics the analytic treatment of geometry became even more fruitful than the axiomatic one.

§ 2. RIEMANNIAN GEOMETRY

Riemann's extension of the concept of space did not start from the axiom of the parallels, but centered around the concept of metric.

Riemann developed further a discovery by Gauss according to which the shape of a curved surface can be characterized by the geometry within the surface. Let us illustrate Gauss' idea as follows. We usually characterize the curvature of the surface of a sphere by its deviation from the plane; if we hold a plane against the sphere it touches only at one point; at all other points the distances between plane and sphere become larger and larger. This description characterizes the curvature of the surface of the sphere "from the outside"; the distances

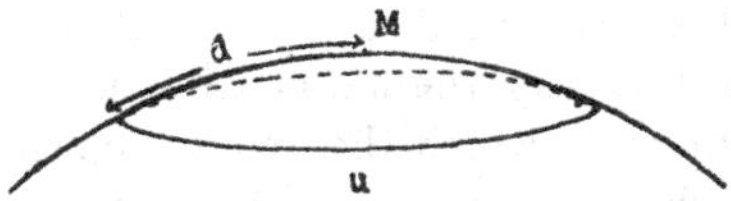

Fig. 1. Circumference and diameter of a circle on the surface of a sphere.

between the plane and the surface of the sphere lie outside the surface and the decision about the curvature has to make use of the third dimension, which alone establishes the difference between curved and straight. Is it possible to determine the curvature of the surface of the sphere without taking outside measurements? Is it meaningful to distinguish the curved surface from the plane within two dimensions? Gauss showed that such a distinction is indeed possible. If we were to pursue "practical geometry" on the sphere, by surveying, for instance, with small measuring rods, we should find out very soon that we were living on a curved surface. For the ratio of circumference u and diameter d of a circle we would obtain a number smaller than $\pi = 3.14...$ as is shown in Fig. 1. Since we stay on the surface all the time, we would not measure the "real diameter" which cuts through the inner part of the sphere, but the "curved diameter" which lies on the surface of the sphere and is longer. This diameter divided into the circumference results in a number smaller than π. Nevertheless, it is meaningful to call the point M "the center of the circle on the surface of the sphere" because it has the same distance from every point of the circle; that we find ourselves on a sphere is noticed by means of the deviation of the ratio from π. In this way we obtain a *geometry of a spherical surface* which is distinguished from the ordinary geometry by the fact that different metrical relations hold for this kind of geometry. In addition to the change in the ratio between circumference and diameter of a circle, an especially important feature is that the sum of the angles of a triangle on a sphere is greater than 180°.

It is remarkable that this generalization of plane geometry to surface geometry is identical with that generalization of geometry which originated from the analysis of the axiom of the parallels. The leading role which has been ascribed to the axiom of the parallels in the course of the development of geometrical axiomatics cannot be justified from a purely axiomatic point of view; the construction of non-Euclidean geometries could have been based equally well upon the elimination of other axioms. It was perhaps due to an intuitive feeling for theoretical fruitfulness that the criticism always centered around the axiom of the parallels. For in this way the axiomatic basis was created for that extension of geometry in which the metric appears as an independent variable.[1] Once the significance of the metric as the characteristic feature of the plane has been recognized from the viewpoint of Gauss' plane theory, it is easy to point out, conversely, its connection with the axiom of the parallels. The property of the straight line of being the shortest connection between two points can be transferred to curved surfaces, and leads to the concept of *straightest line*; on the surface of the sphere the great circles play the role of the shortest line of connection, and on this surface their significance is analogous to that of the straight lines on the plane. Yet while the great circles as "straight lines" share their most important property with those of the plane, they are distinct from the latter with respect to the axiom of the parallels: all great circles of the sphere intersect and therefore there are no parallels among these "straight lines". Here we encounter the second possibility of a denial (cf. § 1) of the axiom of the parallels which excludes the existence of parallels. If this idea is carried through, and all axioms are formulated on the understanding that by "straight lines" are meant the great circles of the sphere and by "plane" is meant the surface of the sphere, it turns out that this system of elements satisfies a system of axioms within two dimensions which is nearly identical in all of its statements with the axiomatic system of Euclidean geometry; *the only exception is the formulation of the axiom of the parallels.*[1] The geometry of the spherical surface can

[1] Cf. p. 148f about the connection of the axiom of the parallels with the metric.

be viewed as the realization of a two-dimensional non-Euclidean geometry: *the denial of the axiom of the parallels singles out that generalization of geometry which occurs in the transition from the plane to the curved surface.*

Once this result has been recognized for two-dimensional structures, a new kind of insight is gained into the corresponding problem of several dimensions by means of a combination of the two different points of departure. The axiomatic development of non-Euclidean geometry had already been achieved for three-dimensional structures and therefore constituted an extension of three-dimensional space analogous to the relation of the plane to the curved surface. Although Euclidean space contains curved surfaces, it does not embody the degree of logical generalization that characterizes the surfaces; it can realize only the Euclidean axiom of the parallels, not the axioms contradicting the latter. This fact suggests a concept of space which contains the plane Euclidean space as a special case, but includes all non-Euclidean spaces too. Such a concept of space in three dimensions is analogous to the concept of surface in two dimensions; it has the same relation to Euclidean space as a surface has to the plane.

On the basis of these ideas Riemann could give so generalized a definition to the concept of space that it includes not only Euclidean space but also Lobatschewsky's space as special cases. According to Riemann, space is merely a three-dimensional manifold; the question is left open which axiomatic systems will hold for it. Riemann showed that it is not necessary to develop an axiomatic system in order to find the different types of space; it is more convenient to use an analytic procedure analogous to the method developed by Gauss for the theory of surfaces. The geometry of space is established in terms of six functions, the *metrical coefficients of the line element,* which must be given [2] as a function of the coordinates; the manipulation of these functions replaces geometrical considerations, and all properties of geometry can be expressed analytically. This procedure can be

[1] It is evident, in considering the spherical surface, that two great circles will intersect in two points; hence, the denial of the axiom that two straight lines can intersect in only one point is involved. For if all of the axioms of Euclidean geometry except the parallel axiom are unchanged it is possible to prove there is at least one parallel. In the treatment of the spherical surface, however, we have seen that this theorem does not hold. This theorem depends upon the axiom that straight lines intersect in only one point; hence its denial removes the inconsistency.

[2] Cf. the more detailed presentation in § 39.

likened to the method in elementary analytic geometry which establishes an equivalence between a formula with two or three variables and a curve or a surface. The imagination is thus given conceptual support that carries it to new discoveries. In analogy to the auxiliary concept of the curvature of a surface, which is measured by the reciprocal product of the main radii of curvature, Riemann introduced the auxiliary concept of *curvature of space*, which is a much more complicated mathematical structure. Euclidean space, then, has a curvature of degree zero in analogy to the plane, which is a surface of zero curvature. Euclidean space occupies the middle ground between the spaces of positive and negative curvatures: it can be shown that this classification corresponds to the three possible forms of the axiom of the parallels. In the space of positive curvature *no* parallel to a given straight line exists; in the space of zero curvature *one* parallel exists; in the space of negative curvature *more than one* parallel exists. In general, the curvature of space may vary from point to point in a manner similar to the point to point variation in the curvature of a surface; but the spaces of *constant curvature* have a special significance. The space of constant negative curvature is that of Bolyai-Lobatschewsky; the space of constant zero curvature is the Euclidean space; the space of constant positive curvature is called spherical, because it is the three-dimensional analogue to the surface of the sphere. The analytical method of Riemann has led to the discovery of more types of space than the synthetic method of Bolyai and Lobatschewsky, which led only to certain spaces of constant curvature. Modern mathematics treats all these types of space on equal terms and develops and manipulates their properties as easily as those of Euclidean geometry.

§ 3. THE PROBLEM OF PHYSICAL GEOMETRY

Let us now return to the question asked at the end of § 1. The geometry of physical space had to be recognized as an empirical problem; it is the task of physics to single out the *actual* space, i.e., physical space, among the *possible* types of space. It can decide this question only by empirical means: but how should it proceed?

The method for this investigation is given by Riemann's mathematical procedure: the decision must be brought about by *practical measurements in space.* In a similar way as the inhabitants of a spherical surface can find out its spherical character by taking measurements, just as we humans found out about the spherical shape of our earth which we cannot view from the outside, it must be possible to find out, by means of measurements, the geometry of the space in which we live. *There is a geodetic method of measuring space analogous to the method of measuring the surface of the earth.* However, it would be rash to make this assertion without further qualification. For a clearer understanding of the problem we must once more return to the example of the plane.

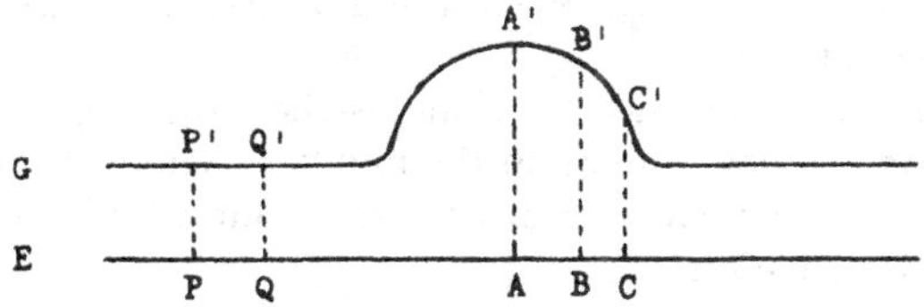

Fig. 2. Projection of a non-Euclidean geometry on a plane.

Let us imagine (Fig. 2) a big hemisphere made of glass which merges gradually into a huge glass plane; it looks like a surface *G* consisting of a plane with a hump. Human beings climbing around on this surface would be able to determine its shape by geometrical measurements. They would very soon know that their surface is plane in the outer domains but that it has a hemispherical hump in the middle; they would arrive at this knowledge by noting the differences between their measurements and two-dimensional Euclidean geometry.

An opaque plane *E* is located below the surface *G* parallel to its plane part. Vertical light rays strike it from above, casting shadows of all objects on the glass surface upon the plane. Every measuring rod which the *G*-people are using throws a shadow upon the plane; we would say that these shadows suffer deformations in the middle area. The *G*-people would measure the distances $A'B'$ and $B'C'$ as equal in length, but the corresponding distances of their shadows AB and BC would be called unequal.

Let us assume that the plane *E* is also inhabited by human beings and let us add another strange assumption. On the plane a mysterious force varies the length of all measuring rods moved about in that plane, so that they are always equal in length to the corresponding shadows projected from the surface G. Not only the measuring rods, however, but all objects, such as all the other measuring instruments and the bodies of the people themselves, are affected in the same way; these people, therefore, cannot directly perceive this change. What kind of measurements would the E-people obtain? In the outer areas of the plane nothing would be changed, since the distance $P'Q'$ would be projected in equal length on PQ. But the middle area which lies below the glass hemisphere would not furnish the usual measurements. Obviously the same results would be obtained as those found in the middle region by the G-people. Assume that the two worlds do not know anything about each other, and that there is no outside observer able to look at the surface E—what would the E-people assert about the shape of their surface?

They would certainly say the same as the G-people, i.e., that they live on a plane having a hump in the middle. They would not notice the deformation of their measuring rods. But why would they not notice this deformation?

We can easily imagine it to be caused by a physical factor, for instance by a source of heat under the plane E, the effects of which are concentrated in the middle area. It expands the measuring rods so that they become too long when they approach A. Geometrical relations similar to those we assumed would be realized; the distances CB and BA would be covered by the same measuring rod and heat would be the mysterious force we imagined.

But could the E-people discover this force? Before we answer this question we have to formulate it more precisely. If the E-people knew that their surface is really a plane, they could, of course, notice the force by the discrepancy between their observed geometry and Euclidean plane geometry. The question, therefore, should read: how can the effect of the force be discovered if the nature of the geometry is not known? Or better still: how can the force be detected if the nature of the geometry may not be used as an indicator?

If heat were the affecting force, *direct* indications of its presence could be found which would not make use of geometry as an *indirect* method. The E-people would discover the heat by means of their sense of temperature. But they would be able to demonstrate the heat expansion independently of this sensation, due to the fact that heat affects different materials in different ways. Thus the E-people would obtain one geometry when using copper measuring rods and another when using wooden measuring rods. In this way they would notice the existence of a *force*. Indeed, direct evidence for the presence of heat is based on the fact that it affects different materials in *different* ways. The fact that the difference in temperature at the points A and C is demonstrable by the help of a thermometer is based on this phenomenon; if the mercury did not expand more than the glass tube and the scale of the thermometer, the instrument would show the same reading at all temperatures. Even the physiological effect of heat upon the human body depends upon differences in the reactions of different nerve endings to heat stimuli.

Heat as a force can thus be demonstrated directly. The forces, however, which we introduced in our example, cannot be demonstrated directly. They have two properties:

(a) They affect all materials in the same way.
(b) There are no insulating walls.

We have discussed the first property, but the second one is also necessary if the deformation is to be taken as a purely metrical one; it will be presented at greater length in § 5. For the sake of completeness the definition of the insulating wall may be added here: it is a covering made of any kind of material which does not act upon the enclosed object with forces having property *a*. Let us call the forces which have the properties *a* and *b* *universal forces*; all other forces are called *differential forces*. Then it can be said that only differential forces, but not universal forces, are directly demonstrable.

After these considerations, what can be stated about the shape of the surfaces E and G? G has been described as a surface with a hump and E as a plane which appears to have a hump. By what right do we make this assertion? The measuring results are the same on both surfaces. If we restrict ourselves to these results, we may just as well say that G is the surface with the "illusion" of the hump and E the surface with the "real" hump. Or perhaps both surfaces have a hump. In our example we assumed from the beginning that E was a plane and G a surface with a hump. By what right do we distinguish between E and G? Does E differ in any respect from G?

These considerations raise a strange question. We began by asking for the actual geometry of a real surface. We end with the question: Is it meaningful to assert geometrical differences with respect to real surfaces? This peculiar indeterminacy of the problem of physical geometry is an indication that something was omitted in the formulation of the problem. We forgot that a unique answer can only be found if the question has been stated exhaustively. Evidently some assumption is missing. Since the determination of geometry depends on the question whether or not two distances are really equal in length (the distances AB and BC in Fig. 2), we have to know beforehand what it means to say that two distances are "really equal." Is *really equal* a meaningful concept? We have seen that it is impossible to settle this question if we admit universal forces. Is it, then, permissible to ask the question?

Let us therefore inquire into the epistemological assumptions of measurement. For this purpose an indispensable concept, which has so far been overlooked by philosophy, must be introduced. The concept of a *coordinative definition* is essential for the solution of our problem.

§ 4. COORDINATIVE DEFINITIONS

Defining usually means reducing a concept to other concepts. In physics, as in all other fields of inquiry, wide use is made of this procedure. There is a second kind of definition, however, which is also employed and which derives from the fact that physics, in contradistinction to mathematics, deals with real objects. Physical knowledge is characterized by the fact that concepts are not only defined by other concepts, but are also coordinated to real objects. This coordination cannot be replaced by an explanation of meanings, it simply states that *this concept* is coordinated to *this particular thing*. In general this coordination is not arbitrary. Since the concepts are interconnected by testable relations, the coordination may be verified as true or false, if the requirement of uniqueness is added, i.e., the rule that the same concept must always denote the same object. The method of physics consists in establishing the uniqueness of this coordination, as Schlick[1] has clearly shown. But certain preliminary coordinations must be determined before the method of coordination can be carried through any further; these first coordinations are therefore definitions which we shall call *coordinative definitions*. They are *arbitrary*, like all definitions; on their choice depends the conceptual system which develops with the progress of science.

Wherever metrical relations are to be established, the use of coordinative definitions is conspicuous. If a distance is to be measured,

[1] M. Schlick, *Allgemeine Erkenntnislehre*, Springer, Berlin 1918, Ziff. 10.

the unit of length has to be determined beforehand by definition. This definition is a coordinative definition. Here the duality of conceptual definition and coordinative definition can easily be seen. We can define only by means of other concepts what we mean by a unit; for instance: "A unit is a distance which, when transported along another distance, supplies the measure of this distance." But this statement does not say anything about the size of the unit, which can only be established by reference to a physically given length such as the standard meter in Paris. The same consideration holds for other definitions of units. If the definition reads, for instance: "A meter is the forty-millionth part of the circumference of the earth," this circumference is the physical length to which the definition refers by means of the insertion of some further concepts. And if the wave-length of cadmium light is chosen as a unit, cadmium light is the physical phenomenon to which the definition is related. It will be noticed in this example that the method of coordinating a unit to a physical object may be very complicated. So far nobody has seen a wave-length; only certain phenomena have been observed which are theoretically related to it, such as the light and dark bands resulting from interference. In principle, a unit of length can be defined in terms of an observation that does not include any metrical relations, such as "that wave-length which occurs when light has a certain redness." In this case a sample of this red color would have to be kept in Paris in place of the standard meter. The characteristic feature of this method is the coordination of a concept to a physical object. These considerations explain the term "coordinative definition." If the definition is used for measurements, as in the case of the unit of length, it is a *metrical* coordinative definition.

The philosophical significance of the theory of relativity consists in the fact that it has demonstrated the necessity for metrical coordinative definitions in several places where empirical relations had previously been assumed. It is not always as obvious as in the case of the unit of length that a coordinative definition is required before any measurements can be made, and pseudo-problems arise if we look for truth where definitions are needed. The word "relativity" is intended to express the fact that the results of the measurements depend upon the choice of the coordinative definitions. It will be shown presently how this idea affects the solution of the problem of geometry.

After this solution of the problem of the unit of length, the next step leads to the comparison of two units of lengths at different locations.

If the measuring rod is laid down, its length is compared only to that part of a body, say a wall, which it covers at the moment. If two separate parts of the wall are to be compared, the measuring rod will have to be transported. It is assumed that the measuring rod does not change during the transport. It is fundamentally impossible, however, to detect such a change if it is produced by universal forces. Assume two measuring rods which are equal in length. They are transported by different paths to a distant place; there again they are aid down side by side and found equal in length. Does this procedure prove that they did not change on the way? Such an assumption would be incorrect. The only observable fact is that the two measuring rods are always equal in length at the place where they are compared to each other. But it is impossible to know whether on the way the two rods expand or contract. An expansion that affects all bodies in the same way is not observable because a direct comparison of measuring rods at different places is impossible.

An optical comparison, for instance by measuring the angular perspective of each rod with a theodolite, cannot help either. The experiment makes use of light rays and the interpretation of the measurement of the lengths depends on assumptions about the propagation of light.

The problem does not concern a matter of *cognition* but of *definition.* There is no way of knowing whether a measuring rod retains its length when it is transported to another place; a statement of this kind can only be introduced by a definition. For this purpose a coordinative definition is to be used, because two physical objects distant from each other are *defined* as equal in length. It is not the *concept* equality of length which is to be defined, but a *real object* corresponding to it is to be pointed out. A physical structure is coordinated to the concept equality of length, just as the standard meter is coordinated to the concept unit of length.

This analysis reveals how definitions and empirical statements are interconnected. As explained above, it is an observational fact, formulated in an empirical statement, that two measuring rods which are shown to be equal in length by local comparison made at a certain space point will be found equal in length by local comparison at every other space point, whether they have been transported along the same or different paths. When we add to this empirical fact the definition that the rods shall be called equal in length when they are at *different places*, we do not make an inference from the observed fact; the addition constitutes an independent convention. There is, however, a certain relation between the two. The physical fact makes the convention unique, i.e., independent of the path of transportation. The statement about the uniqueness of the convention is therefore empirically verifiable and not a matter of choice. One can say that the factual relations holding for a local comparison of rods, though they do not require the definition of congruence in terms of transported rods, make this definition admissible. Definitions that are not unique are inadmissible in a scientific system.

This consideration can only mean that the factual relations may be used for the simple definition of congruence where any rigid measuring rod establishes the congruence. If the factual relations did not hold, a special definition of the unit of length would have to be given for every space point. Not only at Paris, but also at every other place a rod having the length of a "meter" would have to be displayed, and all these arbitrarily chosen rods would be called equal in length by definition. The requirement of uniformity would be satisfied by carrying around a measuring rod selected at random for the purpose of making copies and displaying these as the unit. If two of these copies were transported and compared locally, they would be different in length, but this fact would not "falsify" the definition. In such a world it would become very obvious that the concept of congruence is a definition; but we, in our simple world, are also permitted to choose a definition of congruence that does not correspond to the actual behavior of rigid rods. Thus we could arrange measuring rods, which in the ordinary sense are called equal in length, and, laying them end to end, call the second rod half as long as the first, the third one a third, etc. Such a definition would complicate all measurements, but epistemologically it is equivalent to the ordinary definition, which calls the rods equal in length. In this statement we make use of the fact that the definition of a unit at only one space point does not render general measurements possible. For the general case the definition of the unit has to be given in advance as a function of the place (and also of the time).[1] *It is again a matter of fact that our world admits of a simple definition of congruence because of the factual relations holding for the behavior of rigid rods; but this fact does not deprive the simple definition of its definitional character.*

The great significance of the realization that congruence is a matter of definition lies in the fact that by its help the epistemological problem

[1] Cf. § 39 and § 46.

of geometry is solved. The determination of the geometry of a certain structure depends on the definition of congruence. In the example of the surface E the question arose whether or not the distances AB and BC are equal; in the first case the surface E will have the same geometrical form as the surface G, in the second case it will be a plane. The answer to this question can now be given in terms of the foregoing analysis: whether $AB = BC$ is not a matter of cognition but of definition. If in E the congruence of widely separated distances is defined in such a way that $AB = BC$, E will be a surface with a hump in the middle; if the definition reads differently, E will be a plane. *The geometrical form of a body is no absolute datum of experience, but depends on a preceding coordinative definition*; depending on the definition, the same structure may be called a plane, or a sphere, or a curved surface. Just as the measure of the height of a tower does not constitute an absolute number, but depends on the choice of the unit of length, or as the height of a mountain is only defined when the zero level above which the measurements are to be taken is indicated, geometrical shape is determined only after a preceding definition. This requirement holds for the three-dimensional domain in the same way as it does for the two-dimensional. While in the two-dimensional case the observed non-Euclidean geometry can be interpreted as the geometry of a curved surface in a Euclidean three-dimensional space, we arrive at a three-dimensional non-Euclidean geometry when we measure a three-dimensional structure. A simple consideration will clarify this point. Let us choose as our coordinative definition that of practical surveying, i.e., let us define rigid measuring rods as congruent, when transported. If under these conditions a large circle, say with a radius of 100 meters, is measured on the surface of the earth, a very exact measurement will furnish a number smaller than $\pi = 3.14...$ for the relation of circumference and diameter. This result is due to the curvature of the surface of the earth, which prevents us from measuring the real diameter going through the earth below the curved surface. In this case it would be possible to use the third dimension. If we add the third dimension, however, the situation becomes different. Imagine a large sphere made of tin which is supported on the inside by rigid iron beams; on the sphere and upon the iron scaffold people are climbing around who are measuring circumference and diameter at different points with the same measuring rods they used for the two-dimensional case. If this time the measuring result deviates from π, we must accept a three-dimensional non-Euclidean geometry which can no longer be interpreted as the curvature of a surface in three-dimensional Euclidean space. We obtain this result because the coordinative definition of congruence was chosen as indicated above. A different geometry would have been obtained, if we had used, for instance, the coordinative definition of the earlier example, in which we called the measuring rod half its length after putting it down twice, a third its length after putting it down three times, etc. The question of the geometry of real space, therefore, cannot be answered before the coordinative definition is given which establishes the congruence for this space.

We are now left with the problem: which coordinative definition should be used for physical space? Since we need a geometry, a decision has to be made for a definition of congruence. Although we must do so, we should never forget that we deal with an arbitrary decision that is neither true nor false. Thus the geometry of physical space is not an immediate result of experience, but depends on the choice of the coordinative definition.

In this connection we shall look for the most adequate definition, i.e., one which has the advantage of logical simplicity and requires the least possible change in the results of science. The sciences have implicitly employed such a coordinative definition all the time, though not always consciously; the results based upon this definition will be developed further in our analysis. It can be assumed that the definition hitherto employed possesses certain practical advantages justifying its use. In the discussion about the definition of congruence by means of rigid rods, this coordinative definition has already been indicated. The investigation is not complete, however, because an exact definition of the *rigid body* is still missing.

§ 5. RIGID BODIES

Experience tells us that physical objects assume different states. Solid bodies have an advantage over liquid ones because they change their shape and size only very little when affected by outside forces. They seem, therefore, to be useful for the definition of congruence. However, if the result of the previous considerations is kept in mind, this relative stability is no ground on which to base a preference for solid bodies. As was explained, the form and size of an object depends on the coordinative definition of congruence; if the solid body is used for the coordinative definition, the statement that it does not change its shape must not be regarded as a cognitive statement. It can only be a definition: we define the shape of the solid body as unchangeable. But how can the solid body be defined? In other words, if the physical state of *being solid* were defined differently, under what conditions would the solid body be called *rigid*? If the conservation of shape is not permissible as a criterion, what criteria may be used?

The problem becomes more complicated because we cannot solve it by merely pointing to certain real objects. Although the standard meter in Paris was cited previously as the prototype of such a definition, this account was a somewhat schematic abstraction. Actually no object is the perfect realization of the rigid body of physics; it must be remembered that such an object may be influenced by many physical forces. Only after several corrections have been made, for example, for the influence of temperature and elasticity, is the resulting length of the object regarded as adequate for the coordinative definition of the comparison of lengths. The standard meter in Paris would not be accepted as the definition of the unit of length, if it were not protected from influences of temperature, etc., by being kept in a vault. If an earthquake should ever throw it out of this vault and deform its diameter, nobody would want to retain it as the prototype of the meter; everybody would agree that the standard meter would no longer be a meter. But what kind of definition is this, if the definition may some day be called false? Does the concept of coordinative definition become meaningless?

The answer is: it does not become meaningless, but, as we shall see, its application is logically very complicated. The restrictions that affect the arbitrariness of the coordinative definition have two sources. One restriction lies in the demand that the obtained metric retain certain older physical results, especially those of the "physics of daily life." Nobody could object on logical grounds if the bent rod would be taken as the definition of the unit of length; but then we must accept the consequence that our house, our body, the whole world has become larger. Relative to the coordinative definition it has, indeed, become larger, but such an interpretation does not correspond to our habitual thinking. We prefer an interpretation of changes involving an individual thing on the one side and the rest of the world on the other side that confines the change to the small object. The theory of motion uses the same idea; the fly crawling around in the moving train is called "moving" relative to the train, and the train is called "moving" relative to the earth. Provided that we realize that such a description cannot be justified on logical grounds, we can employ it without hesitation because it is more convenient; yet it must not be regarded as "more true" than any other description. We must not assume that a deformation of the standard meter by an earthquake is equivalent to a change in any absolute sense; actually it is only a change in the *difference* in size between the rod and the rest of the world. There is, of course, no objection to the use of such restrictions on coordinative definitions, because their only effect is an adaptation of the scientific definition to those of everyday life.

These restrictions are more numerous than might be anticipated offhand. Geometrical concepts abound in our daily life. We call the floor and ceiling plane, the corners of our rooms rectangular, a taut string straight. It is clear that these terms can only be definitions and have nothing to do with cognition, as one might at first believe. But by means of these definitions we have arrived at a very simple physics of everyday life. It would logically be permissible to define the taut string as curved, but then we would have to introduce a complicated field of force which pulls the string to the side and prevents it from adjusting itself to the shortest line in spite of the elastic tension, comparable to a stretched chain bending under the influence of gravity; such a convention would complicate physics unnecessarily. However, this is the only objection that can be raised against this description; the statement that a taut string is straight is not empirical but only a more convenient definition.

On the other hand, these restrictions do not constitute strict rules; they merely confine coordinative definitions to certain limits. Direct observation is inexact and we admit the possibility of small inaccuracies of observation. Scientifically speaking nobody will deny that the floor is a little curved, or that a tightened string sags slightly. Such a statement would mean that science does not really use the floor and the string but other physical objects as standards for its coordinative definition, and that, compared to these other things, small deviations occur. The physics of everyday life furnishes only limits for coordinative definitions; it does not intend to establish them strictly.

For everyday physics this strictness is not possible, and the task of scientific physics is therefore to give a strict formulation of the coordinative definition within these limits. This aim of precision is the reason for the important role played by correction factors and supplementary forces in the measurement of lengths. The principle according to which the strict definition is achieved must now be investigated more closely. What is the rigid body of physics? It must be defined strictly without the use of the concept of change in size.

For this purpose the concepts *rigid* and *solid* must be distinguished. Solid bodies are bodies having a certain physical state which can be defined ostensively; it differs from the liquid and gaseous state in a number of observable ways. The solid body can be defined without the use of the concept of change in size. Rigid bodies, however, are those bodies that constitute the physical part in the coordinative definition of congruence and that by definition do not change their size when transported. By the use of the concept *solid body* a definition of the concept *rigid body* can be given that does not employ congruence.

Definition: *Rigid bodies are solid bodies which are not affected by differential forces, or concerning which the influence of differential forces has been eliminated by corrections; universal forces are disregarded.*

This definition will be discussed presently. Let us first deal with the last clause. May we simply neglect universal forces? But we do not neglect them: we merely set the universal forces equal to zero by definition. Without such a rule the rigid body cannot be defined. Since there is no demonstrable difference produced by universal forces, the conception that the transported measuring rod is deformed by such forces can always be defended. No object is rigid relative to universal forces.

This idea corresponds to the usual method of physics. All forces occurring in physics are differential forces in the sense of our definition. The terms "physical forces" and "differential forces" will therefore be used interchangeably in the following sections.

We must still discuss the first part of the definition of the rigid body. Again we shall use the method of the physicist. However, we shall avoid the vicious circle of defining the absence of exterior forces by an absence of change of shape. Since universal forces were eliminated by definition and exterior forces are always demonstrable by differential effects, the conservation of shape is defined inversely through the lack of exterior forces.

This rule needs an addition. It is not possible, even by computations, to eliminate exterior forces completely; small effects evade experimental observation and the definition supplies an ideal limit that can only be approximated. The method of approximation must therefore be discussed. Solid bodies possess considerable interior forces or tensions. According to the usual conception, these forces account for resistance against change of shape; but conversely, in our epistemological construction we can base the definition of negligible change of shape upon the occurrence of these interior forces and tensions. *Change of shape is called small if the exterior forces are small relative to the interior forces.* The more nearly this condition is realized, the more rigid is the body; but only at the unattainable limit where the exterior forces disappear relative to the interior forces would the rigid body be realized in the strict sense.

The definition of the rigid body depends on the definition of a *closed* system. Here lies the difficulty of the problem. Two critical points have been evaded by our definition. In the first place, a closed system can never be strictly realized; therefore, a transition to a limit must be given that permits us to call a system "closed to a certain degree of exactness." This transition to a limit is obtained through the relation between interior and exterior forces, which can be made very small by means of technical manipulation. Without the consideration of interior forces, however, the concept of a closed system could not be determined, because there is always a certain connection with the environment, and it is necessary to name the other magnitudes relative to which the exterior forces are small. It is, therefore, a necessary condition for a closed system to contain interior forces, and even in the transition to infinitesimal closed systems, exterior forces must vanish in a higher order than the interior ones. The second difficulty in the definition of closed systems lies in the possible existence of forces not demonstrable by differential measurements because they affect all indicators in the same way. Physical forces in the sense of our definition can be excluded by adequate protection; but if there exist forces which penetrate all insulating walls (property *b*, p. 13) there are no closed systems. As universal forces they were set equal to zero by definition and as such eliminated. Without such a rule a closed system cannot be defined.

This definition of the rigid body is not explicitly given in the literature of physics, but it is that definition on which the whole system of physics is based. With a different definition physical laws would generally change; this follows from the fact that in the dimensions of the fundamental physical magnitudes, such as force and energy, the concept of length occurs; thus the values of these magnitudes depend on the definition of congruence. It must not be argued, however, that conversely the "truth" of our definition of congruence can be inferred from the truth of physical laws. The truth of the physical laws can only be asserted under the assumption of a definition of congruence; *the laws are true relative to the definition of congruence* by means of rigid bodies. The following example will illustrate this point: if a rubber band were used as the definition of congruence without any indication of its state of tension, the energy of closed systems would in general not be constant, since the measure of the energy would vary as a function of the rubber band. The kinetic energy would change, for instance, because the velocity of the body under consideration would vary with the changes in the rubber band. The law of conservation of energy would be replaced by a law stating the dependence of the energy of closed systems on the state of the rubber band. But this law would be just as true as the law of the conservation of energy. The disadvantage would consist only in the fact that the biography of the rubber band would have to be included in all physical laws. It is one of the most important facts of natural science that it is possible to establish physical laws free from such complications; the significance of the rigid body is based on it.

§ 6. THE DISTINCTION BETWEEN UNIVERSAL AND DIFFERENTIAL FORCES

Our definition of the rigid body is based mainly upon the distinction between universal and differential forces. When we used heat as a differential force in our example above, we could show that a direct proof of physical forces is possible because of the difference of their effects on different materials. This idea must be elaborated further. The thermometer works because mercury and glass do not have the same coefficient of expansion. But can differences in temperature be demonstrated only by differences between the reactions of various materials to heat?

When we recall how the coefficient of expansion of a rod is measured

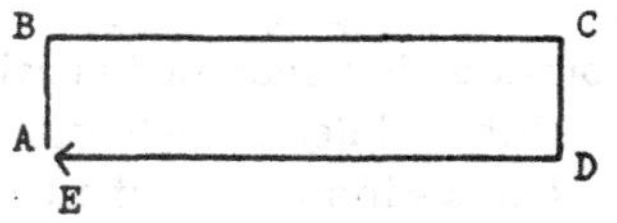

Fig. 3. Sketch of an apparatus for the measurement of heat expansion.

in practice, another possibility suggests itself. For this measurement a device is used as shown in Fig. 3. The distance *ED* corresponds to the rod to be measured. The end *D* is pressed firmly against the support; the end E can move freely. Before the rod is heated, its length is equal to the distance AD. The heat is applied only to ED, while BC is kept at its initial temperature; thus the interval AD remains constant, while ED changes its length. E will move to the left beyond A. The influence of the heat is observable because A and E no longer coincide. This effect is observable, even if the whole apparatus consists of the same material. Imagine a copper wire bent in the rectangular shape of Fig. 3; the two ends of the wire meet in A and E. Such a device would be a "thermometer", because it would be possible to observe a change in temperature by the disappearance of the coincidence between A and E. Here the force is measured by an indicator made of only one material.

Such a device can serve quite generally to demonstrate the presence of forces; the indicator of the force will always react when the field of force is not homogeneous, i.e., if it affects the different parts of the wire in different ways. The field of force may fill the space continuously; if a measurement is to be taken in the field of heat, a complete insulation of the rod DE from the support, i.e., a discontinuity of the field of temperature, is not necessary for the qualitative demonstration of the expansion.

The indicator can have yet another form that makes its operation even more obvious. Imagine a circle made of wire with a diameter

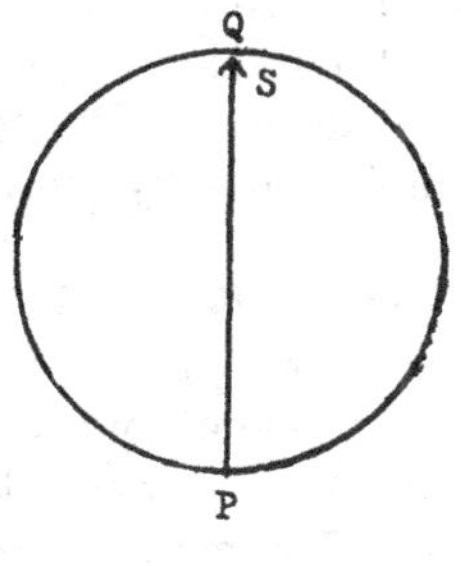

Fig. 4. Sketch of an indicator for the geometrical curvature.

of the same material (Fig. 4). At P this diameter is fastened to the circle, at Q the point S touches the ring, so that there is a coincidence between Q and S. Such an apparatus will also demonstrate the existence of higher temperature in the middle of the circle, for then Q

will not coincide with S. The device can be used for other purposes, too. If it were moved along an egg-shaped surface so that the wire were everywhere in contact with the surface, Q would no longer coincide with S. The indicator points out directly the curvature of the surface by comparing circumference and diameter of a circle. Applied to surfaces of variable curvature, such as an egg-shaped surface, the indicator will register the curvature.

Here we have an indicator of geometrical relations, and we notice that evidence of a field of heat is furnished by a geometrical method. From the change of the geometry we infer the presence of the field of heat. We did not exclude the possibility of this inference; we must, however, analyze the question why, in this case, we go beyond the observation of the geometry, and infer a deforming force. Again we answer that the different reactions of the different kinds of material lead to this inference. In a field of heat the points Q and S on a copper indicator would be shifted in a different way than on an indicator made of iron wire; on an egg-shaped surface both would show the same differences. Thus the only distinguishing characteristic of a field of heat is the fact that it causes different effects on different materials. But we could very well imagine that the coefficients of heat expansion of all materials might be equal—then no difference would exist between a field of heat and the geometry of space. It would be permissible to say that in the neighborhood of a warm body the geometry is changed just as (according to Einstein) space is curved in the neighborhood of a large mass. Nothing could prevent us from carrying through this conception consistently. We do not adopt this procedure because we would then obtain a special geometry for copper, another one for iron, etc.; we avoid these complications by means of the definition of the rigid body.

Although we introduced the differential effects on different materials as an indication of physical forces, a test for forces is not necessarily bound to this difference. A field of force can be demonstrated by the help of one material alone, if the device is large enough to include inhomogeneities of the field. This method, however, will always provide indirect evidence only, because the observed change could equally well be interpreted as a change of geometry. That the change is interpreted as being due to a force can only be based on the unequal effect of the force on different materials. This criterion tells us what should be interpreted as physical deformation and what as geometry of space. The geometry of space, too, can be demonstrated objectively since its physical effects are observable. The distinction between universal and differential forces merely classifies the phenomena as belonging in geometry or in physics.

A remark may be added concerning the treatment these questions receive in the literature. The forces which we called universal are often characterized as forces *preserving coincidences*; all objects are assumed to be deformed in such a way that the spatial relations of adjacent bodies remain unchanged. In this context belongs the assumption that overnight all things enlarge to the same extent, or that the size of transported objects is uniformly affected by their position. Helmholtz' parable of the spherical mirror comparing the world outside and inside the mirror is also of this kind; [1] if our world were to be so distorted as to correspond to the geometrical relations of the mirror images, we would not notice it, because all coincidences would be preserved. It has been correctly said that such forces are not demonstrable, and it has been correctly inferred that they have to be set equal to zero by definition if the question concerning the structure of space is to be meaningful. It follows from the foregoing considerations that this is a *necessary* but not a *sufficient* condition. Forces *destroying coincidences* must also be set equal to zero, if they satisfy the properties of the universal forces mentioned on p. 13; only then is the problem of geometry uniquely determined. Our concept of universal force is thus more general and contains the concept of the coincidence-preserving force as a special case. It should not be said, therefore, that universal forces are not demonstrable; this holds only for forces which preserve coincidences. Fig. 4, however, is an example of an indicator showing universal forces which destroy coincidences (in this case the coincidence QS).

We can define such forces as equal to zero because a force is no absolute datum. When does a force *exist*? By force we understand something which is responsible for a *geometrical change*. If a measuring rod is shorter at one point than at another, we interpret this contraction as the effect of a force. The existence of a force is therefore dependent on the coordinative definition of geometry. If we say: actually a geometry G applies, but we measure a geometry G', we define at the same time a force F which causes the difference between G and G'. The geometry G constitutes the zero point for the magnitude of a force. If we find that there result several geometries G' according

[1] H. v. Helmholtz, *Schriften zur Erkenntnistheorie*, ed. by Hertz and Schlick, Springer, Berlin 1921, p. 19.

as the material of the measuring instrument varies, F is a differential force; in this case we gauge the effect of F upon the different materials in such a way that all G' can be reduced to a common G. If we find, however, that there is only one G' for all materials, F is a universal force. In this case we can renounce the distinction between G and G', i.e., we can identify the zero point with G', thus setting F equal to zero. This is the result that our definition of the rigid body achieves.

§ 7. TECHNICAL IMPOSSIBILITY AND LOGICAL IMPOSSIBILITY

In the following section a criticism will be discussed which has been made against our theory of coordinative definitions. It has been objected that we base the arbitrariness in the choice of the definition on the impossibility of making measurements. Although it is admitted that certain differences cannot be *verified* by measurement, we should not infer from this fact that they do not *exist*. If we had no means of discovering the shape of surface E in Fig. 2 (p. 11) it would still be meaningful to ask what shape the surface has; although the possibility of making measurements is dependent on our human abilities, the objective fact is independent of them. Thus we are accused of having confused *subjective inability* with *objective indeterminacy*.

There are, indeed, many cases where physics is unable to make measurements. Does this mean that the magnitude to be measured does not exist? It is impossible, for instance, to determine exactly the number of molecules in a cubic centimeter of air; we can say with a high degree of certainty that we shall never succeed in counting every individual molecule. But can we infer that this number does not exist? On the contrary, we must say that there will always be an integer which denotes this quantity exactly. The mistake of the theory of relativity is supposed to consist in the fact that it confuses the *impossibility of making measurements* with *objective indeterminacy*.

Whoever makes this objection overlooks an important distinction. There is an impossibility of making measurements which is due to the limitation of our technical means; I shall call it *technical impossibility*. In addition, there is a *logical impossibility* of measuring. Even if we had a perfect experimental technique, we should not be able to avoid this logical impossibility. It is logically impossible to determine whether the standard meter in Paris is really a meter. The highest refinement of our geodetic instruments does not teach us anything about this problem, because the meter cannot be defined in absolute terms. This is the reason why the measuring rod in Paris is called the definition of a meter. It is arbitrarily defined as the unit, and the question whether it really represents this unit has lost its meaning. The same considerations hold for a comparison of units at distant places. Here we are not dealing with technical limitations, but with a logical impossibility. The impossibility of a determination of the shape of a surface, if universal forces are admitted, is not due to a deficiency of our instruments, *but is the consequence of an unprecise question.* The question concerning the shape of the surface has no precise formulation, unless it is preceded by a coordinative definition of congruence. What is to be understood by "the shape of a real surface"? Whatever experiments and measurements I make, they will never furnish a unique indication of the shape of the surface. If universal forces are admitted, the measurements may be interpreted in such a way that many different shapes of surfaces are compatible with the same observations. There is one definition which closes the logical gap and tells us which interpretations of our observations must be eliminated: this task is performed by the coordinative definition. It gives a precise meaning to the question of the shape of the real surface and makes a unique answer possible, just as a question about length has a unique meaning only when the unit of measurement is given. It is not a technical failure that prevents us from determining the shape of a surface without a coordinative definition of congruence, but a logical impossibility that has nothing to do with the limitations of human abilities.

The situation will be further clarified if we compare the last example with the case of the indeterminacy of the number of molecules in a given cubic centimeter of air. This number is precisely defined and it is only due to human imperfection that we cannot determine it exactly. But in this case an approximation is possible which will increase with increasing perfection of our technical instruments. When we are faced with a logical impossibility there are no approximations. We cannot decide approximately whether the surface E of Fig. 2 (p. 11) is a plane, or a surface with a hemispherical hump in the middle; there is no defined limit which the measurement could approach. Furthermore, once the coordinative definition is given, the technical impossibility of an exact measurement remains. Even our definition of the rigid body does not permit a strict determination of the structure of space; all our measurements will still contain some degree of inexactness which a progressive technique will gradually reduce but never overcome.

§ 8. THE RELATIVITY OF GEOMETRY

With regard to the problem of geometry we have come to realize that the question which geometry holds for physical space must be decided by measurements, i.e., empirically. Furthermore, this decision is dependent on the assumption of an arbitrary coordinative definition of the comparison of length. Against this conception arguments have been set forth which endeavor to retain Euclidean geometry for physical space under any circumstances and thus give it a preference among all other geometries. On the basis of our results we can discuss these arguments; our analysis will lead to the relativity of geometry.

One of the arguments maintains it is a mistake to believe that the choice of the coordinative definition is a matter left to our discretion. The measurements of geometry as carried through in practice presuppose quite complicated measuring instruments such as the theodolite; therefore these measurements cannot be evaluated without a theory of the measuring instruments. The theory of the measuring instruments, however, presupposes the validity of Euclidean geometry and it constitutes a contradiction to infer a non-Euclidean geometry from the results.

This objection can be met in the following way. Our conception permits us to start with the assumption that Euclidean geometry holds for physical space. Under certain conditions, however, we obtain the result that there exists a universal force F that deforms all measuring instruments in the same way. However, we can invert the interpretation: we can set F equal to zero by definition and correct in turn the theory of our measuring instruments. We are able to proceed in this manner because a transformation of all measurements from one geometry into another is possible and involves no difficulties. It is correct to say that all measurements must be preceded by a definition; we expressed this fact by the indispensability of the coordinative definition. The mistake of the objection consists in the belief that this definition cannot be changed afterwards. Just as we can measure the temperature with a Fahrenheit thermometer and then convert the results into Celsius, measurements can be started under the assumption of Euclidean geometry and later converted into non-Euclidean measurements. There is no logical objection to this procedure.

In practice the method is much simpler. It turns out that the non-Euclidean geometry obtained under our coordinative definition of the rigid body deviates quantitatively only very little from Euclidean geometry when small areas are concerned. In this connection "small area" means "on the order of the size of the earth"; deviations from Euclidean geometry can be noticed only in astronomic dimensions. In practice, therefore, it is not necessary to correct the theory of the measuring instruments afterwards, because these corrections lie within the errors of observation. The following method of inference is permissible: we can prove by the assumption that Euclidean geometry holds for small areas that in astronomic dimensions a non-Euclidean geometry holds which merges infinitesimally into Euclidean geometry. No logical objection can be advanced against this method, which is characteristic of the train of thought in modern physics. It is carried through in practice for astronomic measurements designed to confirm Einstein's theory of gravitation.

The objection is connected with the *a priori* theory of space that goes back to Kant and today is represented in various forms. Not only Kantians and Neo-Kantians attempt to maintain the *a priori* character of geometry: the tendency is also pronounced in philosophical schools which in other respects are not Kantian. It is not my intention to give a critical analysis of Kant's philosophy in the present book. In the course of the discussion of the theory of relativity, it has become evident that the philosophy of Kant has been subject to so many interpretations by his disciples that it can no longer serve as a sharply defined basis for present day epistemological analysis. Such an analysis would clarify less the *epistemological* question of the structure of space than the *historical* question of the meaning and content of Kant's system. The author has presented his own views on this problem in another publication; [1] the present investigation is aimed at philosophical clarification and will not concern itself with historical questions. Therefore, I shall select only those arguments of Kant's theory of space, the refutation of which will further our understanding of the problem. Although in my opinion the essential part of Kant's theory will thereby be covered, I do not claim a historically complete evaluation of it in this book.

The ideas expressed in the preceding considerations attempted to establish Euclidean geometry as *epistemologically a priori*; we found that this *a priori* cannot be maintained and that Euclidean geometry

[1] H. Reichenbach, *Relativitätstheorie und Erkenntnis a priori*, Springer, Berlin, 1920.

is not an indispensable presupposition of knowledge. We turn now to the idea of the *visual a priori*; this Kantian doctrine bases the preference for Euclidean geometry upon the existence of a certain manner in which we visualize space.

The theory contends that an innate property of the human mind, the ability of visualization, demands that we adhere to Euclidean geometry. In the same way as a certain self-evidence compels us to believe the laws of arithmetic, a visual self-evidence compels us to believe in the validity of Euclidean geometry. It can be shown that this self-evidence is not based on logical grounds. Since mathematics furnishes a proof that the construction of non-Euclidean geometries does not lead to contradictions, no *logical* self-evidence can be claimed for Euclidean geometry. This is the reason why the self-evidence of Euclidean geometry has sometimes been derived, in Kantian fashion, from the human ability of visualization conceived as a source of knowledge.

Everybody has a more or less clear notion of what is understood by visualization. If we draw two points on a piece of paper, connect them by a straight line and add a curved connecting line, we "see" that the straight line is shorter than the curved line. We even claim to be certain that the straight line is shorter than any other line connecting the two points. We say this without being able to prove it by measurements, because it is impossible for us to draw and measure all the lines. The power of imagination compelling us to make this assertion is called the ability of *visualization*. Similarly, the Euclidean axiom of the parallels seems to be visually necessary. It remains for us to investigate this human quality and its significance for the problem of space.

The analysis will be carried through in two steps. Let us first assume it is correct to say that a special ability of visualization exists, and that Euclidean geometry is distinguished from all other geometries by the fact that it can easily be visualized. The question arises: what consequences does this assumption have for physical space? Only after this question has been answered can the assumption itself be tested. The second step of our analysis will therefore consist in the inquiry whether a special ability of visualization exists (§ 9–§ 11).

Let us turn to the first question, which has to be reformulated in order to relate it clearly to the epistemological problem.

Mathematics proves that every geometry of the Riemannian kind can be mapped upon another one of the same kind. In the language of physics this means the following:

Theorem θ: "Given a geometry G' to which the measuring instruments conform, we can imagine a universal force F which affects the instruments in such a way that the actual geometry is an arbitrary geometry G, while the observed deviation from G is due to a universal deformation of the measuring instruments."[1]

No epistemological objection can be made against the correctness of theorem θ. Is the visual *a priori* compatible with it?

Offhand we must say yes. Since the Euclidean geometry G_0 belongs to the geometries of the Riemannian kind, it follows from theorem θ that it is always possible to carry through the visually preferred geometry for physical space. Thus we have proved that we can always satisfy the requirement of visualization.

But something more is proved by theorem θ which does not fit very well into the theory of the visual *a priori*. The theorem asserts that Euclidean geometry is not preferable on epistemological grounds. Theorem θ shows all geometries to be equivalent; it formulates the *principle of the relativity of geometry*. It follows that it is meaningless to speak about one geometry as the *true* geometry. We obtain a statement about physical reality only if in addition to the geometry G of the space its universal field of force F is specified. Only the combination

$$G+F$$

is a testable statement.

We can now understand the significance of a decision for Euclidean geometry on the basis of a visual *a priori*. The decision means only the choice of a specific coordinative definition. In our definition of the rigid body we set $F = 0$; the statement about the resulting G is then a univocal description of reality. This definition means that in "$G+F$" the second factor is zero. The visual *a priori*, however, sets $G = G_0$. But then the empirical component in the results of measurements is represented by the determination of F; only through the combination

$$G_0+F$$

are the properties of space exhaustively described.

There is nothing wrong with a coordinative definition established on

[1] Generally the force F is a tensor. If $g'_{\mu\nu}$ are the metrical coefficients of the geometry G' and $g_{\mu\nu}$ those of G, the potentials $F_{\mu\nu}$ of the force F are given by

$$g'_{\mu\nu}+F_{\mu\nu} = g_{\mu\nu} \qquad \mu\nu = 1, 2, 3$$

The measuring rods furnish directly the $g'_{\mu\nu}$; the $F_{\mu\nu}$ are the "correction factors" by which the $g'_{\mu\nu}$ are corrected so that $g_{\mu\nu}$ results. The universal force F influencing the measuring rod is usually dependent on the orientation of the measuring rod. About the mathematical limitation of theorem θ cf. §12.

the requirement that a certain kind of geometry is to result from the measurements. We ourselves renounced the simplest form of the coordinative definition, which consists in pointing to a measuring rod; instead we chose a much more complicated coordinative definition in terms of our distinction between universal and differential forces. A coordinative definition can also be introduced by the prescription what the result of the measurements is to be. "The comparison of length is to be performed in such a way that Euclidean geometry will be the result"—this stipulation is a possible form of a coordinative definition. It may be compared to the definition of the meter in terms of the circumference of the earth: "The unit is to be chosen in such a way that 40 million times this length will be equal to the circumference of the earth."

Although it may be admitted that Euclidean geometry is unique in that it can be easily visualized, the theory of the visual *a priori* does not disprove the theory of the relativity of geometry and of the necessity for coordinative definitions of the comparison of length. On the contrary, it is only this theory that can state precisely the epistemological function of visualization: the possibility of visualization is a ground for subjective preference of one particular coordinative definition. But the occurrence of visualization does not imply anything about the space of real objects.

In this connection another argument in support of the preference for Euclidean geometry is frequently adduced. To be sure, this argument is not related to the problem of visualization, but like the visual *a priori* it attributes a specific epistemological position to Euclidean geometry; therefore we shall consider it here. It is maintained that Euclidean geometry is the *simplest* geometry, and hence physics must choose the coordinative definition $G = G_0$ rather than the coordinative definition $F = 0$. This point of view can be answered as follows: physics is not concerned with the question which *geometry* is simpler, but with the question which *coordinative definition* is simpler. It seems that the coordinative definition $F = 0$ is simpler, because then the expression $G+F$ reduces to G. But even this result is not essential, since in this case simplicity is not a criterion for truth. Simplicity certainly plays an important part in physics, even as a criterion for choosing between physical hypotheses. The significance of simplicity as a means to knowledge will have to be carefully examined in connection with the problem of induction, which does not fall within the scope of this book.

Geometry is concerned solely with the simplicity of a *definition*, and therefore the problem of empirical significance does not arise. It is a mistake to say that Euclidean geometry is "more true" than Einstein's geometry or vice versa, because it leads to simpler metrical relations. We said that Einstein's geometry leads to simpler relations because in it $F = 0$. But we can no more say that Einstein's geometry is "truer" than Euclidean geometry, than we can say that the meter is a "truer" unit of length than the yard. The simpler system is always preferable; the advantage of meters and centimeters over yards and feet is only a matter of economy and has no bearing upon reality. *Properties of reality are discovered only by a combination of the results of measurement with the underlying coordinative definition.* Thus it is a characterization of objective reality that (according to Einstein) a three-dimensional non-Euclidean geometry results in the neighborhood of heavenly bodies, if we define the comparison of length by transported rigid rods. But only the *combination* of the two statements has objective significance. The same state of affairs can therefore be described in different ways. In our example it could just as well be said that in the neighborhood of a heavenly body a universal field of force exists which affects all measuring rods, while the geometry is Euclidean. Both combinations of statements are equally true, as can be seen from the fact that one can be transformed into the other. Similarly, it is just as true to say that the circumference of the earth is 40 million meters as to say that it is 40 thousand kilometers. The significance of this simplicity should not be exaggerated; this kind of simplicity, which we call *descriptive simplicity*, has nothing to do with truth.

Taken alone, the statement that a certain geometry holds for space is therefore meaningless. It acquires meaning only if we add the coordinative definition used in the comparison of widely separated lengths. The same rule holds for the geometrical shape of bodies. The sentence "The earth is a sphere" is an incomplete statement, and resembles the statement "This room is seven units long." Both statements say something about objective states of affairs only if the assumed coordinative definitions are added, and both statements must be changed if other coordinative definitions are used. These considerations indicate what is meant by *relativity of geometry*.

This conception of the problem of geometry is essentially the result of the work of Riemann, Helmholtz, and Poincaré and is known as *conventionalism*. While Riemann prepared the way for an application of geometry to physical reality by his mathematical formulation of the concept of space, Helmholtz laid the philosophical foundations. In particular, he recognized the connection of the problem of geometry with that of rigid bodies and interpreted correctly the possibility of a visual representation of non-Euclidean spaces (cf. p. 63). It is his merit, furthermore, to have clearly stated that Kant's theory of space is untenable in view of recent mathematical developments.[1] Helmholtz' epistemological lectures must therefore be regarded as the source of modern philosophical knowledge of space.[2] It is Einstein's achievement to have applied the theory of the relativity of geometry to physics. The surprising result was the fact that the world is non-Euclidean, as the theorists of relativity are wont to say; in our language this means: if $F = 0$, the geometry G becomes non-Euclidean. This outcome had not been anticipated, and Helmholtz and Poincaré still believed that the geometry obtained could not be proved to be different from Euclidean geometry. Only Einstein's theory of gravitation predicted the non-Euclidean result which was confirmed by astronomical observations. The deviations from Euclidean geometry, however, are very small and not observable in everyday life.

Unfortunately, the philosophical discussion of conventionalism, misled by its ill-fitting name, did not always present the epistemological aspect of the problem with sufficient clarity.[3] From conventionalism the consequence was derived that it is impossible to make an objective

[1] The antithesis Kant-Helmholtz has been interpreted by Neo-Kantians (in particular by Riehl, *Kantstudien* 9, p. 261f., less plainly by Görland, *Natorp-Festschrift*, p. 94f) not as a contradiction but as a misunderstanding of Kant by Helmholtz. The same argument has been advanced by Neo-Kantians recently with respect to Einstein's theory. This conception is due to an underestimation of the differences between the points of view, and it would be in the interest of a general clarification if the patent contradiction between the only possible modern philosophy of space and Kant were admitted. Such an admission avoids the danger of an interpretation of Kant's philosophy too vague to retain any concrete content. The author presented his ideas on the subject in "Der gegenwärtige Stand der Relativitätsdiskussion," *Logos* X, 1922, section III, p. 341. Cf. also p. 31. (The English translation of this paper will be included in a forthcoming volume of *Selected Essays* by Hans Reichenbach, to be published by Routledge and Kegan Paul, London.)

[2] Cf. the new edition by Hertz and Schlick, *Helmholtz' Erkenntnistheoretische Schriften*, Berlin 1921.

[3] This is also true of the expositions by Poincaré, to whom we owe the designation of the geometrical axioms as conventions (*Science and Hypothesis*, Dover Publications, Inc. 1952, p. 50) and whose merit it is to have spread the awareness of the definitional character of congruence to a wider audience. He overlooks the possibility of making objective statements about real space in spite of the relativity of geometry and deems it impossible to "discover in geometric empiricism a rational meaning" (*op. cit.*, p. 79). Cf. § 44.

statement about the geometry of physical space, and that we are dealing with subjective arbitrariness only; the concept of geometry of real space was called meaningless. This is a misunderstanding. Although the statement about the geometry is based upon certain arbitrary definitions, the statement itself does not become arbitrary: once the definitions have been formulated, it is determined through objective reality alone which is the actual geometry. Let us use our previous example: although we can define the scale of temperature arbitrarily, the indication of the temperature of a physical object does not become a subjective matter. By selecting a certain scale we can stipulate a certain arbitrary number of degrees of heat for the respective body, but this indication has an objective meaning as soon as the coordinative definition of the scale is added. On the contrary, it is the significance of coordinative definitions to lend an objective meaning to physical measurements. As long as it was not noticed at what points of the metrical system arbitrary definitions occur, all measuring results were undetermined; only by discovering the points of arbitrariness, by identifying them as such and by classifying them as definitions can we obtain objective measuring results in physics. *The objective character of the physical statement is thus shifted to a statement about relations.* A statement about the boiling point of water is no longer regarded as an absolute statement, but as a statement about a relation between the boiling water and the length of the column of mercury. There exists a similar objective statement about the geometry of real space: *it is a statement about a relation between the universe and rigid rods.* The geometry chosen to characterize this relation is only a mode of speech; however, our awareness of the relativity of geometry enables us to formulate the objective character of a statement about the geometry of the physical world as a statement about relations. In this sense we are permitted to speak of *physical geometry.* The description of nature is not stripped of arbitrariness by naive absolutism, but only by recognition and formulation of the points of arbitrariness. The only path to objective knowledge leads through conscious awareness of the role that subjectivity plays in our methods of research.

§ 9. THE VISUALIZATION OF EUCLIDEAN GEOMETRY

With the result of the foregoing section in mind we turn now to the second question essential to the theory of the visual *a priori* of Euclidean geometry: Is it true that Euclidean geometry is the only geometry which can be visualized? If Euclidean geometry is distinguished by being easily visualized, this fact does not add anything to our knowledge of physical space. But does Euclidean geometry have this distinction? Let us analyze this question.

The investigation will have to proceed in two directions. On the one hand, we must inquire whether other geometries can be visualized; this question is usually taken to be the main one. On the other hand —and this question will be the subject matter of the present section— we must find out what visualization of Euclidean geometry means, and to what extent Euclidean geometry can be visualized. The visualization of Euclidean geometry should by no means be taken for granted; on the contrary, we shall carefully examine this assertion that has so frequently been maintained by philosophers.

At the very start of our investigation we encounter a difficulty. As soon as we try to give a more precise formulation to the experience of visualization, we are in the midst of psychological experiments: we try to analyze geometrical images. As a consequence we have those philosophers against us who maintain that the problem of visualization concerns not psychological but philosophical issues. In particular, this conception is represented by Neo-Kantians who assert that Kant's pure intuition is not a psychological phenomenon. These objections must not deter us. It has always turned out that such "border skirmishes" do not help in the study of the problems; sometimes it is much harder to make a decision about the classification of a problem than about its solution. We shall therefore disregard the objections and try to find the right method of analyzing the experience of geometrical visualization.

Two characteristic features stand out: it is of the nature of visualization that it reproduces the particular object in the form of an image. When we attempt to visualize an object, for instance a triangle, blurred images emerge in our mind that are obviously connected with previous perceptions. We may imagine a white triangle on a blackboard, or a triangle drawn by pencil on a white sheet of paper; but the image always appears somewhat schematic. Individual details occur only when we concentrate on vivid reproductions of perceived triangles; we suddenly see that the lines of chalk marking the triangle have a certain width and that they are composed of individual particles of chalk. The schematic triangle, however, is also determined by previous perceptions. It is not flaming red on a blue background—such triangles are rarely seen; the schematic triangle resembles a triangle drawn by chalk or pencil much more closely than such a product of the imagination. It is of medium shape and not a degenerate triangle with rarely-perceived unusual relations between the sides. In this connection a peculiar indeterminateness can be observed: it is difficult for us to estimate the size of the angle at the top. If we want a more precise statement, we must concentrate much harder; only then can we reproduce the triangle vividly enough in our imagination to estimate the size of the angle. It is unnecessary to analyze these phenomena more closely, since they vary greatly with the individual person. Let us call the function of visualization described above its *image-producing* function.

Apart from the image-producing function—and this constitutes the second characteristic feature of visualization—it has a *normative function*. Visualization is not completely arbitrary. We use visualization in order to discover geometrical relations. I have a triangle and a straight line intersecting one of the sides of the triangle; if sufficiently prolonged, will the straight line also intersect another side of the triangle? Visualization says "yes." It simply demands this answer and I can do nothing about it. I try to turn the straight line in my imagination; I see that the line can be managed to intersect one *or* the other side of the triangle, but I am unable to prevent it from intersecting *either* side. It is simply impossible. This normative function is the philosophically more important component of visualization; it is the cause of the philosophical controversies about the epistemological significance of visualization. Kant's *synthetic a priori of pure intuition* springs from the normative function of visualization. It is this function that tends to single out Euclidean geometry from all the others; it seems to compel us to regard Euclid's axiom of the parallels as unquestionably true.

What is the source of the normative function and the compulsion which it expresses? It seems that the source is the image-producing function, because the image-producing function is a necessary condition of the effectiveness of the normative function. Only after we imagine the triangle and the intersecting line do we "see" that the law mentioned above holds. There are cases where we contemplate a problem for some time without being able to solve it, until we succeed in producing a clear image; we then read the desired law from the image. How many diagonals can be drawn from one corner of a pentagon? We are not able to answer this question immediately; at first we make some vain efforts and hope to find the answer so to speak offhand without the use of the image-producing function. We do not succeed. We must first mobilize our will; suddenly the pentagon is visualized with its characteristic asymmetry, one corner at the top, one side at the bottom. Now we draw two diagonals from the bottom left corner to the two corners at right; more are evidently impossible. This vivid image lends certainty to our answer.

The vividness of the picture and the certainty of the judgment are enhanced if we actually draw the pentagon. We see the result immediately, and it is no longer necessary to strain our will. But obviously this facilitation is the only function of the drawing. The perception of the drawing does not play the role of an empirical discovery; the situation is not comparable to the performing of an experiment as a consequence of which we expect perception to supply an answer. If we combine two unknown salt solutions, we expect the perception to tell us whether or not a precipitate is the result. The drawing, however, has a different function. We do not read results from it, but read them into it. We also reserve the privilege of correcting the drawing, because drawings are not always reliable. Suppose we draw a pentagon with all of its diagonals and try to count them. If we draw a pentagon very carefully we shall succeed. But it may easily happen that one corner is drawn too flat, and thus is overlooked when we draw some of the diagonals; in this case we must correct the drawing. Apparently we do not trust our perception but follow an inner compulsion; the perception of the drawing serves only to facilitate the image-producing function. The detour by way of the drawing makes the final statement more certain.

These considerations suggest that the normative function does not have its origin in the image-producing function. We correct not only the drawing but also the images themselves by means of the normative function. Sometimes the image-producing function provides us with a false image. I put the problem of the pentagon mentioned above to a person untrained in mathematics. I immediately got the rash answer "five." He was evidently in the phase of speaking offhand. Then followed a "no, one moment." Now the image-producing function was employed, and after some reflection came the answer "three." Here the image-producing function had evidently furnished the wrong result. A "no" followed and after some moments the correct answer "two." The normative function had intervened and corrected the images. It is not the case that we simply wait for images that will dictate the result to us. On the contrary, the images are subject to a directive, and if they do not correspond to it, they will not be accepted.

This directive is stronger than is usually suspected, and it works even behind the scenes. It not only refers to conditions indicated in the problem but also adds some tacit conditions. We considered the theorem that a straight line intersecting one side of a triangle must also intersect another side of the triangle. Is this true? By no means; I can imagine a straight line descending in space and not situated in the same plane as the triangle; in this case it intersects one side only. This answer is certainly trivial—but often we do not notice how much we restrict a problem by tacit assumptions. Some parlor games make use of this unawareness. Three matches are laid on the table in the shape of a triangle; the problem is to form four triangles by adding three more matches. Rarely somebody conceives the idea of arranging the three matches spatially on top of the triangle lying on the table so that a tetrahedron results. And in this category belongs the story of the egg of Columbus: note what conditions you impose upon your imagination; then many an "impossible" turns out to be an "impossible under such and such conditions."

The humorous aspect of the examples just mentioned is due to the fact that it would be quite easy to eliminate the tacit assumptions; the questions, however, are asked in such a way that they suggest them. Since the matches are put on the table it is suggested that the puzzle concerns a problem of the plane. Apart from the particular aspect of this problem, such experiences furnish the key for quite a few difficulties of geometrical visualization. Rather late in the history of mathematics the *analysis situs* was discovered, which led to certain peculiarities of visualization. Does there exist a surface having only one side? Visualization suggests a prompt "no." But every student of a lecture on topology has taken a strip of paper, and twisted once around itself, pasted it together in form of a ring; this paper surface has indeed only one side. After we have seen such a model, our ability to visualize has increased. Or again: a closed curve is drawn on a surface; is it possible to draw on the surface a line of any shape that connects a point of the surface situated on one side of the curve with a point of the surface situated on the other side of the curve, without intersecting the curve? Visualization again answers "no," but only because the image-producing function shows a plane. Therefore, we attempt to solve the problem in the plane, where it is impossible. Mathematics has shown that there are surfaces of different topological properties where not all closed curves divide the surface into separate areas. We can very well imagine such surfaces; or more precisely: we can direct the image-producing function of visualization so that it will furnish elements for which the desired property holds. If occasionally we are answered "impossible" by the images of our visualization, we must first inquire to what extent tacit assumptions are contained in the elements furnished by the image-producing function; the presence of these assumptions may prevent us from producing an image of the problem under consideration. Only if these assumptions are explicitly stated in the problem is it correctly answered. In the example of the straight line intersecting the side of a triangle, it must be stated explicitly that the straight line must lie in the plane of the triangle. Only then is the problem worded correctly. In this instance it seems to be easy to add the missing assumption later, but it is not always so easy. We do not know, either, whether we tacitly add other assumptions in the theorem of the triangle given above.

In this very example some further conditions play a part. The theorem is correct only if it refers to Euclidean straight lines and to a Euclidean plane. It does not hold for every kind of surface and its straightest lines; the theorem has exceptions on the torus, for instance, (Fig. 1, Plate I). It is not a matter of escaping into the third dimension as above; all lines remain within the same two-dimensional surface. In order to formulate the theorem strictly, the condition must be added that by "straight line" and "plane" the respective phenomena of Euclidean geometry are understood. Only then is it conclusive.

But then there is nothing spectacular about the visual complusion inherent in the theorem, because the visual compulsion does not furnish anything different from the logical compulsion inherent in Euclidean geometry. In the system of Euclidean geometry this theorem is necessary and we discover this fact by logical analysis of the geometry whose axioms include this theorem.[1] Without this theorem the elements of geometry, such as the straight line and the plane, have very different properties. The merit of visualization consists only in the fact that it translates the logical compulsion of Euclidean geometry into a visual compulsion. *The normative function of visualization is revealed as a correlate of the logical compulsion and achieves the same results by means of the elements furnished by the image-producing function as the logical inference does by means of the conceptual elements of thought.* This is the significance of visualization. It seems to be much easier to make logical inferences with the help of visual representation than

[1] This axiom was first formulated by Pasch. Cf. D. Hilbert, *Foundations of Geometry*, The Open Court Publishing Co., Chicago, 1921. p. 6.

by means of abstract concepts. Proofs which the mathematician has found only with great effort—as for instance the theorem that a continuous function assumes every intermediate value between any two of its values—became immediately obvious through visualization. In this gift of visual inference our mind possesses one of the most powerful tools, not only for science but also for everyday life. It is certainly wonderful that such an achievement of visualization is possible; but it is not an achievement outside the frame of logic. The manner in which logical inferences are actually made is strange and obscure and rarely resembles the formal method of logic. But this fact is irrelevant for the problem of the visualization of geometry. We may therefore take it for granted (not subject to an investigation in this connection) that visual processes play a role in logical thinking.

The examples from two-dimensional geometry which we selected are very instructive for our problem. If we speak of a special visual compulsion, we express the idea that more restrictive laws hold for visualization than for logical thinking. It is this idea which Kant meant by his synthetic *a priori* judgments of visualization, and it is also the reaction which aprioristic philosophy had all along to the existence of non-Euclidean geometries: it is possible to construct them logically, but it is impossible to visualize them. According to this conception, visualization admits a narrower selection of geometrical structures than does logic. It turns out that this statement is not true in the two-dimensional domain. The normative function of visualization does not demand more than logic; what is logically consistent can be visualized as far as it does not surpass the precision attainable by visual representation—for the two-dimensional realm this is certainly true. It is not different in the three-dimensional domain. The images which we habitually use are those of Euclidean geometry. No wonder that we derive only Euclidean laws from them. It never happens in the game of chess that the two original bishops belonging to one player stand on squares of the same color. It is, however, possible simply to put them upon two white squares. The statement about the bishops is correct only so long as the rules of the game are adhered to; the law of the oblique path of the bishop logically implies the necessity that the color will remain unchanged. The images by which we visualize geometry are always so adjusted as to correspond to the laws which we read from them; these laws are always implied. The statement that we cannot visualize non-Euclidean geometry must therefore be reformulated: *We cannot visualize non-Euclidean geometry by means of Euclidean elements of visualization.* In this form the result is trivial; what it denies is a logical impossibility. The question must be asked differently: Can we change the image-producing elements in such a way that we can read the laws of non-Euclidean geometry from the new images? Only in such a manner can we attempt a visualization of non-Euclidean geometry. The question will be treated in § 11. We shall have to analyze, in particular, at what point tacit assumptions are smuggled into the images, assumptions that make them Euclidean and cause the normative function to reject non-Euclidean laws.

§ 10. THE LIMITS OF VISUALIZATION

Before taking up the problem of the visualization of non-Euclidean geometry we must consider another fact the significance of which is controversial. It was explained above that the image-producing function, under the directive of the normative function, furnishes images from which logical laws can be read by means of visualization. It must be noticed, however, that there exist limitations for visualization that prevent the production of images from going beyond certain simple relations. We are able to visualize a pentagon but do not succeed with a decagon without drawing it on a piece of paper. A polygon of a thousand sides, even if we see it drawn on a piece of paper, has lost the specific image-character that distinguishes it from a polygon with one thousand-and-four sides. These are the limitations that compel the mathematician to renounce visual methods in favor of analytical ones. Nobody will attempt to count by means of visualization the number of diagonals in a polygon with a thousand sides; we would not even trust the result of counting them in a drawing, but would always prefer the analytical method, which derives the number of diagonals from the number of corners by a simple formula. There are individual differences in the ability of visualizing geometry, but they are restricted to a certain domain; beyond this domain there begins for everybody the larger realm of geometrical figures which can no longer be visualized. It would be a mistake to say that the entire Euclidean geometry can be visualized. Only the elementary geometrical figures can be visualized, that is, realized through images.

Even these elementary geometrical structures can be visualized only within certain limitations depending on the size of the figures. We frequently say that we can visualize the terrestial globe; but this seems to be a mistake. We can visualize a sphere, but not of the dimensions of the earth. What we visualize, when we speak of the possibility of visualizing the earth, is a small sphere which we believe to be similar to the terrestrial globe. We can try to enlarge the image of this sphere; we may perhaps visualize it as big as a balloon or as a mountain—but very soon we must admit failure. It may be objected that there do not exist absolute sizes in our images—but that is not true. A sphere of the size of a balloon has a visual quality different from that of a sphere the size of a tennis-ball, and such a sphere has again a quality different from that of the spherical head of a pin. In our imagination such structures can be distinguished just as well as a triangle and a square. The fact that in actual perception size is a matter of perspective has nothing to do with this ability. If it happens occasionally that we see a distant balloon as small as a tennis-ball because it is seen in the same angular perspective as such a ball, it means only that we transfer the visual quality "tennis ball" to the physical structure "balloon." It would be a mistake to say that this transfer is irrelevant to visualization, because the relations in the large dimensions resemble those in the small dimensions. If small images are used to visualize large figures, it is an indirect method that makes use of extraneous pictures. This detour is possible because of a special property of Euclidean geometry. The system of Euclidean geometry has theorems of similarity. In non-Euclidean geometry there are no such theorems; the angular sum of the triangle, and the relation between circumference and diameter of a circle depend there upon the absolute size of the figure. In non-Euclidean geometry, therefore, the indirect method must be modified; the smaller figures must be imagined as distorted. Such analogies are possible only within certain limitations implied by the problem; it is dangerous to look for visual necessities where there are none. Visualization does not necessarily presuppose that large figures can visually be replaced by small figures of the same shape. This is a visual requirement in Euclidean geometry only, and even in this case the transition to smaller figures does not have the same effect as is achieved by direct perception. The image-producing function is replaced by a conceptual coordinative method smuggling in extraneous pictures.

The same is true for very small structures. Figures of atomic size such as the orbit of an electron can no longer be directly visualized. Here, too, the indirect way of visualizing similar figures in the intermediate dimension is employed.

The limits of exactness of visual images in the intermediate domain is connected with the impossibility of visualizing very small figures.

Although we can visualize a right angle, we cannot distinguish it from an angle of 89° 59′. If we are dealing with a triangle having such an angle, there is no other way but to visualize an angle deviating much more from 90°; only then can we notice the difference from the right angle. Two straight-line segments which, if continued, intersect on the sun, are indistinguishable from parallels on the earth. We possess, therefore, only one visual image for both phenomena. Those who object to this statement forget that the two non-parallel segments which they visualize converge too quickly; again an indirect method has been used by the substitution of an extraneous picture that satisfies the specified conditions.

Finally, the problem of the infinity of space must be mentioned among the limitations of visualization. That this question has been disputed so much and that even antinomies have been constructed in connection with it is due only to the fact that here the image-producing function of visualization fails. We cannot visualize Euclidean space as a whole—this is briefly the meaning of all arguments for or against the infinity of space. For conceptual constructions it is very easy to handle the concept of the infinity of space, and in spite of Kant only that proof is correct which infers the infinity of Euclidean space. We are able to make theoretical statements about space as a whole, such as the statement that space is three-dimensional. But it is impossible to visualize it as a whole—to take it in at one glance just as we take in a sphere or a landscape. Attempts to replace direct visualization by letting the eyes glide over homogeneous domains of space are always makeshift. The infinity of space contains a property of Euclidean geometry that cannot be realized by visualization. We shall come back to this question when dealing with the holistic properties of non-Euclidean spaces, where we shall find analogous phenomena.

The limits of visualization described above certainly exist, and it is a mistake often made in philosophical quarters to neglect them as "purely psychological." The psychological realization is important; a logical inference can be realized psychologically in all its strictness, but the visualization of a right angle or of infinite space is impossible. We are justified, therefore, in inferring from these psychological limitations of visualization that it is not the visualization which is responsible for the rigor of our work, but the logic which we always think into our images. Badly drawn figures, where homologous sides of congruent triangles are evidently different in length, nevertheless enable us to give a strict geometrical proof; in the same way, inexact visual images will permit strict logical inferences, because the logical conditions of congruence are fulfilled. The inferences are possible because the compelling factor is not the visualization but the logical laws implicit in the images. An analysis of visualization of Euclidean geometry must therefore stress the limits of this human ability: the limits prove that the normative character of our images does not spring from visualization but from logic.

It is but another mistake when mathematicians attempt to base a visualization of non-Euclidean geometry on just these limits of visualization.[1] The deviations of physical geometry from Euclidean geometry are so small according to the theory of relativity that they are far below the limit of exactness of visualization, and it would be convenient to say that Einsteinian space can be visualized just as well as Euclidean space, since no difference is noticed within the visual field. Yet this argument begs the question. It is not permissible to base a statement about visualization upon phenomena where visualization fails; there are certainly noticeable differences of visualization like the difference between an angle of 90° and one of 45°. The problem of the ability to visual non-Euclidean geometry centers around the question not whether the negative but whether the positive properties of visualization can be exploited for non-Euclidean geometry. If it is possible to visualize non-Euclidean geometry, it must be possible for a space of strong curvature, perceptible, say within the dimensions of a room, just as well as for Einstein's weakly curved space—otherwise it would not make sense to speak of visualization, and one could only say that we are unable to visualize differences between the two spaces. It is, therefore, not expedient to take the *limits* of visualization as the point of departure for our attempt to make non-Euclidean geometry accessible to visualization. Our only concern with these limits has been to establish the fact that it is not visualization, but logic, which dictates the laws read from the images.

In an earlier section (§ 1) non-Euclidean geometry was made plausible by means of "badly-drawn" figures; now we understand the deeper significance of this device. It is the predominance of logic in visualization which is expressed in this method. It will be our task to extend the method in such a way that we can imagine for non-Euclidean geometry even the possibility of "well-drawn" figures. By means of conceptual thinking, we can pass from visualization to readjusted

[1] This conception is suggested by F. Klein, *Elementarmathematik vom höheren Standpunkte aus,* Vol. II, p. 192f. Springer, Berlin 1925.

visualization. The human mind has the ability to circumvent visual images by the use of abstract concepts, as it were. After this achievement it is able to produce new images. We must not demand that the new images have the same degree of immediacy as the old ones, and that they come into focus with as little effort as the images we are accustomed to. Some constraint and some adjustment of the imagination is indispensable; and so is careful analysis which reveals what is actually visual in so-called visual experience.

§ 11. VISUALIZATION OF NON-EUCLIDEAN GEOMETRY

It has occasionally been maintained by mathematicians who have worked a great deal in non-Euclidean geometry that they can gradually visualize it. The proponents of the theory of relativity argue that the visualization of Euclidean geometry is only the result of habit, and that we could gradually acquire the ability to visualize non-Euclidean geometry. One must not forget, however, that very little is gained by such a statement, because we do not know yet what is to be understood by "visualization of non-Euclidean geometry." The mathematician is inclined to postpone these philosophical questions in favor of the mathematical elaboration of geometry. Nobody denies that the mathematician succeeds in getting accustomed to non-Euclidean concepts that permit fast and effective research for mathematical purposes; but the question remains whether such an imagining of

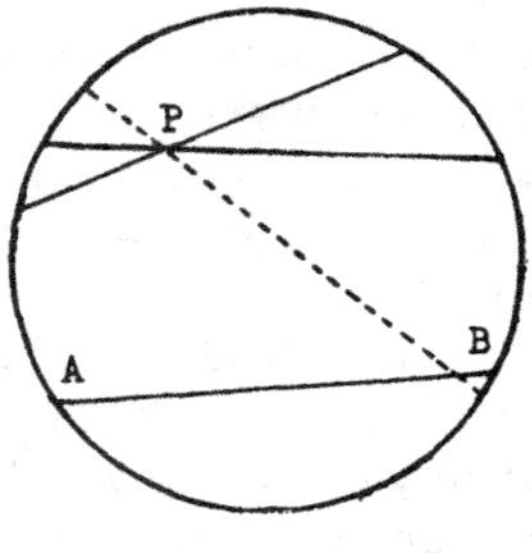

Fig. 5. Klein's model of non-Euclidean geometry.

non-Euclidean relations can be compared to the phenomenon which we call "visualization" in Euclidean geometry.

The mathematicians have developed a procedure that enables us to "visualize" non-Euclidean geometry by means of Euclidean geometry. This method is based upon the mathematical fact that a non-Euclidean geometry can be mapped upon Euclidean space. A well-known example will illustrate this matter. In Fig. 5 a circle is drawn, the interior of which can be used to visualize the geometry of Bolyai-Lobatschewsky. The same relations hold between the *chords* of this circle as between the *straight lines* of Lobatschewsky's geometry, so long as we restrict ourselves to the interior of the circle. Through the point P a (dotted) chord has been drawn intersecting AB; two more chords have been drawn which do not intersect AB. This drawing illustrates the axiom of the parallels of Lobatschewsky's geometry; it means, in the language of this geometry, the existence of several straight lines through one point which do not intersect a given straight line. It may be objected that these straight lines, if sufficiently continued beyond the circle, will intersect the continuation of the straight line AB; they certainly will, but that does not contradict our assertion. We said only that the relations between the chords in the *interior of the circle* are identical with those of Lobatschewsky. Every theorem valid for Lobatschewsky's straight lines is valid for the chords in the interior of the circle. Mathematics can show, furthermore, that Lobatschewsky's entire plane can be mapped upon the interior of this circle. In this mapping, however, the concept of distance between two points of Lobatschewsky's geometry is not coordinated to the concept of distance between two points of Euclidean geometry, but to a mathematical expression which has a very complicated form in Euclidean geometry.[1] This expression has the consequence that equal segments on a chord, in non-Euclidean metric, correspond in Euclidean metric to segments becoming smaller and smaller as we approach the perimeter of the circle, so that an infinite number of such "equal" segments are situated on a chord. By means of this device a theorem of Euclidean geometry is coordinated to every theorem of Lobatschewsky's geometry. Of course, the corresponding theorems have different meanings in the two geometries. For instance, the theorem of Euclidean geometry "There are several chords through one point which

[1] It is the logarithm of the cross-ratio formed by the two points and the points of intersection of the chord with the circle. Cf. H. Weyl, *Space-Time-Matter*, Methuen & Co. Ltd., London, 1922, p. 82.

do not intersect a given chord in the interior of a circle" corresponds to Lobatschewsky's theorem "There are several straight lines through one point which do not intersect a given straight line." A theorem of Lobatschewsky concerning the congruence of triangles would correspond to a theorem of Euclidean geometry which states something about the complicated function that replaces the Euclidean concept of distance.

Are we now able to visualize non-Euclidean geometry? Certainly not. In this example only the Euclidean elements of the theorems can be visualized. Instead of one of Lobatschewsky's theorems, which cannot be visualized, we visualize one of Euclid's theorems, and by means of this detour are enabled to manipulate Lobatschewsky's geometry with greater ease. Lobatschewsky's concepts become abbreviations for more complicated Euclidean relationships; we speak the language of Lobatschewsky but connect with these concepts the visual meaning of Euclidean relations. It is like making intelligible a sentence consisting of a meaningless chance combination of words by coordinating to them new meanings; it cannot be said, however, that thereby the original sentence becomes intelligible.

Philosophers have therefore objected to this method of the mathematicians on the ground that the result is not a visualization of non-Euclidean geometry but a pointing out of a system of relations between elements of Euclidean space, a system which is analogous to the system of non-Euclidean relations. The method was taken to demonstrate that visualization of non-Euclidean geometry is impossible and must therefore be replaced by a mapping upon Euclidean space as the only one that can be visualized. Accordingly, the method of mapping has merely the logical function of proving the consistency of non-Euclidean geometry (cf. § 1). This important result is indisputable, but one refused to regard the method of mapping as a visual representation of non-Euclidean geometry.

This rejection appears to be justified. Therefore, we shall not continue our investigations in connection with this mathematical treatment of the problem, but shall go back to the physical treatment, where the problem of the geometry of real space and of measurement is most conspicuous. In this way we shall approach the problem of visualization more closely, since we are dealing with empirical perceptions, not with conceptual constructions. We shall later come back to the method of mapping described above.

It was explained previously under what circumstances a physicist would decide to call a space "non-Euclidean." He would just keep to the measuring results of rigid rods and let them determine the geometry. Let us visualize his experiences: Assume that the observed geometry is the three-dimensional analogue to Fig. 2 (p. 11). We draw a two-dimensional cross-section (Fig. 6) through space: in the plane

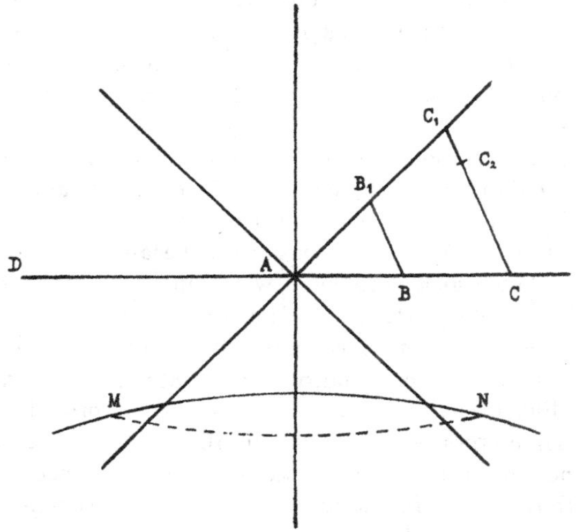

Fig. 6. Cross-section through a non-Euclidean space. In non-Euclidean relations $BB_1 = CC_2$, although $AB = BC$ and $AB_1 = B_1C_1$. The solid line MN is a line of equal distance from DC, the dotted line MN is a straightest line; they do not coincide.

of our drawing it looks similar to the projection of surface G upon surface E of Fig. 2. Fig. 6 may be conceived as the top view of Fig. 2 (cf. however the footnote on p. 52). Imagine meridians drawn from A' (Fig. 2) upon G which merge into radial straight lines in the plane part of G. In E these meridians will appear as straight lines emanating from A, as Fig. 6 shows. We must only add that we imagine Fig. 6 as a cross-section through a similar space so that from A straight lines would emanate in all directions.

In order to describe our observations we assume that the physicist retains Euclidean geometry. He then observes the following. By means of his measuring rod he made $AB = BC = AB_1 = B_1C_1$. The distance BB_1 is as long as a solid rod which he laid down. He now places this rod on the line CC_1. He finds that it goes from C to C_2, whereas C_2C_1 is shorter than the rod. In Euclidean geometry CC_1 must be equal to 2 (BB_1); he will say that the rod has grown under the influence of a field of force F. The force has the effect that the rod becomes longer in tangential position the farther he moves away from A, whereas it remains unchanged in radial position.[1] If he moves still farther away from A, the expansion increases still more, then decreases, until it finally disappears and normal relations obtain. The field of force F is found to be universal, i.e., independent of the material of the measuring rods.

Corresponding relations were described as the result of measurements on the surface E of Fig. 2, where they can be explained as a consequence of the projection. There the transition to non-Euclidean geometry (i.e., a "defining away" of F) was simpler, because we were concerned with a two-dimensional problem: we could visualize surface E as having a hump like surface G. The introduction of non-Euclidean geometry thus does not cause any visual difficulties in the two-dimensional domain. The situation is different in the three-dimensional realm. If Fig. 6 is to represent a cross-section through a space for which the same cross-sections result in all directions originating from A, we cannot interpret this cross-section as a hump because it would conflict with the other cross-sections. The cross-section must therefore remain plane and yet show the geometry of the sphere—this is the requirement contradicting visualization that stops us when we wish to introduce non-Euclidean space. Can we get rid of the contradiction?

We must analyze, therefore, in what respect the interpretation in terms of a hump constitutes a solution for the two-dimensional problem.

[1] He could also say that a contraction occurs in radial position, whereas there is no change in tangential position; the two descriptions would be equivalent. The second one would correspond to the relations of the projection of Fig. 2. We choose the first one because it facilitates the following presentation. In Fig. 6, therefore, the distances AB and BC are equal in length, whereas in Fig. 2, BC is shorter. Fig. 6 is not obtained from the surface G of Fig. 2 by parallel projection but by a projection that preserves the lengths of the meridians, yet expands the parallel circles. In dealing with the field of gravitation in the neighborhood of a mass point, Einstein chooses the second description and speaks of a contraction of the radial measuring rod. (*Ann. d. Phys.* 49, 1916, § 22.) The resulting relations are similar to those described here; the only difference is that the contraction increases rather than decreases when the center is approached. In the center (or even in its vicinity) we find a singularity. It is the geometry of a paraboloid originating from the rotation of a parabola around its vertex tangent (or around straight lines outside the parabola and parallel to the tangent). Cf. L. Flamm, *Physikal Zeitschrift* 17, 1916, p. 438.

How is the curvature of a surface visualized? What we see are usually the differences relative to three-dimensional space rather than the geometrical relations within the surface. The distance between a curved surface and an adjacent plane varies; that is the criterion by which a curvature is visualized. We thus make use of the third dimension in order to visualize the curvature of the two-dimensional manifold. Let us call this kind of curvature the *exterior curvature.* It is well-known that the exterior curvature of a surface can change without a change of the *interior curvature*, i.e., of the geometry of its surface. The zone of a sphere of elastic sheet metal, for instance, can be twisted without being expanded, so that the metal shows a different shape while retaining its inner spherical geometry. (This is not true for the sphere as a whole.) If a piece of paper is fashioned into a cylinder, it will have an exterior curvature, but not an interior curvature, because the rolling does not involve an expansion; the surface of a cylinder has therefore the geometry of the Euclidean plane. What we usually visualize as the curvature of a surface is its exterior curvature. If we were to attempt a similar visualization for a three-dimensional manifold, we would have to imbed it in (at least) a four-dimensional space. Here lies the difficulty of the problem. It seems to be very difficult to impose a new dimension upon visualization. Physically this new dimension would not even be justified, because all physical happenings are confined to the three-dimensional realm. Later we shall investigate more precisely in what sense the number of dimensions is determined by physical occurrences; it is evident, however, that the introduction of a new dimension is useless for our problem, because we cannot make any measurements in the fourth dimension and therefore no criterion analogous to "the distances of the curved surface from a plane" exists. The problem must be solved in the three-dimensional domain, i.e., we must try to visualize the *interior curvature* of the space.

There is another reason why we must stay in three dimensions. Euclidean space can be visualized only in three dimensions. Just as a curved surface can be characterized by its relations to three-dimensional space, a plane can be described in terms of three dimensions. For instance: planes are structures the intersections of which are straight lines. In the same sense in which we speak of the existence of an exterior and an interior curvature, we may speak of the absence of an exterior and interior curvature. If we attempted to describe non-Euclidean space as imbedded in a higher manifold, we could demand the same for Euclidean space. But since we restrict ourselves to a visualization of the absence of the interior curvature, we may also restrict ourselves to a visualization of the interior curvature greater than zero. We wanted to compare our ability to visualize one or the other; therefore we must try to visualize non-Euclidean space in the same manner as Euclidean space, i.e., within the three-dimensional domain.

This insight constitutes the first step on the path to a visualization of non-Euclidean space. We must not demand of an image of non-Euclidean space properties of an image of a surface in three-dimensional space, but should look only for analogues to properties of a surface visualized in two-dimensions, i.e., to the interior curvature of a surface.

What visual experiences in the two-dimensional domain induce us to say that a surface is curved? Only one feature represents the curvature visually. If we look at Fig. 6, the distance CC_2 appears longer than BB_1. If we look at the same measuring rods on the surface G, however, CC_2 appears equal in length to BB_1. The visual experience of the curvature in two dimensions consists in the fact that we judge the relations of congruence differently. It does not matter that the distance CC_1 does not become twice as long as BB_1, in spite of the equality of the sections on the radial lines; we project the congruence differently upon the plane. This insight is the second step on our path to visualize curved space.

We can carry through the same procedure for three dimensions. We can adjust our visualization in such a way that we see the distances BB_1 and CC_2 equal in length. We do not need a hump in the surface; we need only to adjust our conception of congruence. Such an adjustment is permissible because congruence is a matter of *definition.* Even Euclidean congruence, which we often tacitly presuppose, is based on a definition. This definition too is projected by us into space, not discovered in it. The adjustment necessary for a visualization of a curved space consists in projecting congruence differently into three-dimensional space.

We have such a strong visual perception of Euclidean geometry because all our experience with rigid rods constantly teaches us Euclidean congruence. Let us recall the description of the image-producing function of visualization in the foregoing section, where we explained that the manner of visualization is mainly conditioned by previous perceptions. If we wish to change our reaction to Euclidean congruence, we shall have to strain the normative function considerably. It would be different if in daily life we dealt occasionally with rigid bodies that adjusted themselves to non-Euclidean geometry. We can imagine what our experiences would be if we suddenly worked with measuring rods that behaved according to Fig. 6. At first we would have the feeling that objects *changed* when transported, and we would apply the formula G_0+F. After some time we would lose this feeling and no longer perceive any change of the objects when they are transported. Now we would have adjusted our visualization to a geometry G for which $F=0$. When a near-sighted person puts on glasses for the first time, he sees all objects distinctly, but they seem to move as soon as he moves. After a while this feeling wanes, and he has become accustomed to the new way of seeing. We would have the corresponding experience in a non-Euclidean world; the moment we no longer see any change in the transported objects, we have accomplished a visual adjustment.

Let us point to another example of such an adjustment. Automobiles are frequently equipped on the driver's side with a convex mirror showing the lanes in the rear. The untrained person sees the picture in the mirror in a distorted way; moving objects seem to change in shape; but the driver accustomed to the picture no longer has the impression of distortion and change of shape. A corresponding adjustment is made to the many strange perspective relations of our Euclidean environment; children often do not have static pictures: they see a moving train in the size of a toy train and have the impression that the departing train becomes objectively smaller. Neither are they able to identify the static picture of distant congruences with the picture of nearby congruences. Children see the parallel lines of a street as objectively converging, and when they arrive at the end of the street, they cannot understand that this is the same spot which they saw from a distance. *Any adjustment to congruence is a product of habit; the adjustment is made when, during the motion of the objects or of the observer, the change of the picture is experienced as a change in perspective, not as a change in the shape of the objects.*

Whoever has successfully adjusted himself to a different congruence is able to visualize non-Euclidean structures as easily as Euclidean structures and to make inferences concerning them. I should like to use the problem of the parallels as an illustration. There are no parallels in Riemannian space; let us try to visualize this feature. In Fig. 6 the solid line MN is drawn in such a way that it has everywhere a constant distance from the "straightest line" DC. In Euclidean language: MN is curved so that it approaches DC more closely in the middle; a rigid rod put between the lines as a measure of their distance would have a radial direction in the middle of the figure and would not contract, whereas at the sides it would come into a tangential position and would expand. If we adjust our eyes to the other congruence we can very well "see" the distance of the two lines as being the same everywhere. We have to realize only that Euclidean congruence, in spite of its obtrusiveness, is likewise merely a definition which we "see into" the plane of drawing. Furthermore, the solid line *MN* is no straightest line. The dotted line *MN* is the shorter connection between *M* and *N*. In Euclidean geometry this means: a measuring rod transported along the dotted line passes in the middle through zones farther removed from *A* and is therefore longer in these areas; thus it can be laid down less often on the dotted line. If we readjust our eyes to the other congruence, we see clearly that the rod can be laid down less often along the dotted line. Consequently: there are no parallels in this instance; the two descriptions "line of equal distance from a given straight line" and "straightest line" do not apply to the same thing. This is the meaning of the statement that there are no parallels.

In order to explain the drawing of Fig. 6 we have so far used the Euclidean language. This usage, however, is necessary only in the beginning of attempts at such an adjustment. Statements like "the rod is longer here than there" are eventually replaced by statements like "the rod covers here this distance and there this distance," and the corresponding distances are then visualized. Eventually one can forget that from the viewpoint of Euclidean geometry these distances are different in length. We have similar experiences when we learn a new language: in the beginning we can only translate from our native language, and even when we talk, the new words acquire a meaning only because the translation is always present in our mind. Gradually we learn to associate a meaning immediately with the new words, to think in the new language and to express ourselves without the detour through the native language. A similar emancipation from the "native geometry" is experienced with regard to the visualization of non-Euclidean relations.

It is indeed possible to visualize non-Euclidean space by an adjustment of visualization to a different congruence. Euclidean space has thus lost its privileged status. It should not be objected that even this method constitutes a mapping of non-Euclidean relations upon Euclidean space. Space as such is neither Euclidean nor non-

Euclidean, but only a continuous three-dimensional manifold. It becomes Euclidean if a certain definition of congruence is assumed for it; this congruence is mathematically characterized by the fact that the concepts of line of equal distance from a given straight line and of straightest line are coextensive. As long as we adjust our eyes to this congruence, we visualize Euclidean space. If a different definition of congruence is introduced for which the above-mentioned mathematical condition does not hold, space becomes non-Euclidean. We are unable to visualize it so long as we cannot emancipate ourselves from Euclidean congruence; with this restriction non-Euclidean relations can only be mapped upon the visualized Euclidean space. Space will be visualized as non-Euclidean if we succeed in visualizing the new definition of congruence as congruence, i.e., in adjusting our eyes to it. It is, in fact, the result of training the eyes to adjust to the behavior of solid bodies seen in different angular perspectives that enables us to visualize Euclidean congruence. If we readjust the eyes we can similarly visualize non-Euclidean congruence. With this step we have succeeded in visualizing all that can possibly be visualized in Euclidean or non-Euclidean spaces within the three-dimensional domain. We have visualized the interior curvature, since interior curvature is nothing but the deviation from Euclidean congruence.

We can now come back to Klein's method of mapping the Bolyai-Lobatschewsky geometry upon the interior of a circle. The procedure was based on a coordination of non-Euclidean concepts to Euclidean ones. This method, too, reveals itself as a redefinition of congruence and can be visualized by a readjustment to a different congruence. The definition of non-Euclidean congruence is formulated in terms of Euclidean congruence; this kind of definition results from the nature of the model, which is intended to establish a correspondence between the two geometries. The procedure represents the indirect method described above: Euclidean geometry is inserted as an intermediate step for the purpose of rendering possible a visualization of the defined congruence. It is somewhat difficult to forget this detour, but it is possible to conceive the definition of congruence as given directly by the statement "this distance is congruent to that distance." Only as long as Klein's method is identified with a mapping process is the example not a representation of Lobatschewsky's geometry. But it is possible to adjust to the other congruence, i.e., to see as congruent those sections of the chord which, in Euclidean language, become smaller and smaller as the periphery is approached. In this sense

Klein's picture is truly a visual representation of Lobatschewsky's geometry.

Klein's model involves a special difficulty, because Lobatschewsky's infinite space is mapped upon a finite section of Euclidean space. In order to accomplish the visualization, we must "forget" everything outside the circle (in the three-dimensional domain outside the sphere). We must imagine ourselves inside the circle and remember that the periphery cannot be reached in a finite number of steps.

The mathematician is thus correct in saying that he has become accustomed to visualize non-Euclidean geometry by working with it. But we can now analyze the process by which he changes his visualization. He has become accustomed to visualize actually as congruence the definition which originally was given as a function of Euclidean elements, to emancipate himself from the impression of a change, which in the beginning affects everybody working with such congruences, and to project congruence into space in a manner different from the one to which he was previously accustomed. If his realization of the definitional character of congruence is strong enough to direct the image-producing function of his visualization, he will succeed. Our previous thesis was that the apparent impossibility of visualizing non-Euclidean geometry originated from the fact that a contradictory presupposition was smuggled in which demanded that non-Euclidean geometry be visualized in terms of Euclidean elements. This presupposition can now be stated more precisely: *Euclidean congruence is the tacit presupposition that influences the image-producing function of our visualization when it so stubbornly rejects non-Euclidean geometry.* The refusal is certainly justified, because non-Euclidean relations cannot be visualized in terms of Euclidean congruence: such a visualization is logically impossible. Euclidean congruence is the rule presupposed for the game of chess if certain moves seem to be impossible; an alternative is possible only when the rule of the game is changed. After such a change has been achieved, one can read the laws of non-Euclidean geometry, such as the nonexistence of parallels, from one's images in the same way as the untrained person takes Euclidean axioms for granted in his visual images.

§ 12. SPACES WITH NON-EUCLIDEAN TOPOLOGICAL PROPERTIES

The concept *topological* was mentioned previously; let us add a brief explanation. The surfaces of three-dimensional space are distinguished from each other not only by their curvature but also by certain more general properties. A spherical surface, for instance, differs from a plane not only by its roundness but also by its finiteness. Finiteness is a *holistic property*. The sphere as a whole has a character different from that of the plane. A spherical surface made from rubber, such as a balloon, can be twisted so that its geometry changes. We can give it an egg-shaped appearance or press it into the form of a die; but it cannot be distorted in such a way that it will cover a plane. All surfaces obtained by distortion of the rubber sphere possess the same holistic properties; they are closed and finite. The plane as a whole has the property of being open; its straight lines are not closed. This feature is mathematically expressed as follows. Every surface can be mapped upon another one by the coordination of each point of one surface to a point of the other surface, as illustrated by the projection of a shadow-picture by light rays. For surfaces with the same holistic properties it is possible to carry through this transformation *uniquely and continuously* in all points. *Uniquely* means: one and only one point of one surface corresponds to a given point of the other surface, and vice versa. *Continuously* means: neighborhood relations in infinitesimal domains are preserved; no tearing of the surface or shifting of relative positions of points occur at any place. For surfaces with different holistic properties such a transformation can be carried through locally, but there is no single transformation for the whole surface. As an illustration let us take the stereographic projection of the spherical surface (Fig. 9, p. 69). From the north pole P we draw radial lines to project every point of the surface of the sphere upon the horizontal plane. In general this transformation is unique and continuous, although the metrical relations are distorted; for the point P, however, it shows a singularity. Point P is mapped upon the infinite; i.e., no finitely located point of the plane corresponds to it. It can be shown that every transformation possesses a singularity in at least one point. The surface of the sphere is therefore called *topologically different* from the plane. Only a "sphere without a north pole" would be *topologically equivalent* to the plane. This would be a sphere where exactly one mathematical point is excluded, whereas all adjacent points are preserved; such a sphere has a point-shaped hole without a boundary and is no longer a closed surface.

Greater topological differences result when surfaces of a different connectivity are considered. The torus is a doubly-connected surface (Fig. II, Plate 1); it has the shape of a doughnut. Its characteristic feature is that there exist on it closed curves which cannot be contracted to a point. There are no such curves on the plane; if we imagine any closed curve on the plane (Fig. 7), a concentric curve can be drawn inside it, a smaller one inside the second one, and so forth, until the curves are contracted to a point. There are such curves on the torus, but not every curve has this property. The curves drawn in Fig. II (Plate 1) cannot be contracted to a point. If we go from curve 1 to the curves 2 and 3, 2 lies between 1 and 3 just as in Fig. 7 the curve 2

Fig. 7. Curves that can be contracted to a point.

lies between the curves 1 and 3. But if the transition to additional curves is continued on the surface of the torus, we shall finally return to curve 1, since the curves become larger on the other side of the torus in contradistinction to the plane, where the curves are contracted to a point. The statements that 3 lies between 2 and 1, or that 1 lies between 2 and 3, are equivalent, because the order of betweenness[1] does not depend upon the mutual distances of the curves; betweenness is purely a relation of order. That naive visualization believes curve 2 to lie necessarily between 1 and 3 demonstrates the confusion of metrical and order relations and hence the mistake of an uncritical reliance on results of such visualization. Only a conceptual formulation reveals the errors of naive visualization and enables us to correct them. Once the conceptual structure has been understood, visualization follows suit: the relation of betweenness on the torus is undetermined for curves that cannot be contracted to a point, i.e., for three of such curves it is not uniquely determined which of them lies between the other two. (The relations of betweenness for the ring curves on the surface of the torus are the same as those of the points on the periphery of a circle.) This indeterminateness of the betweenness relation has the consequence that such a curve does not divide the surface of the torus into two separate domains; between points to the "right" and

[1] By *betweenness* we shall understand the relation indicated by the word "between."

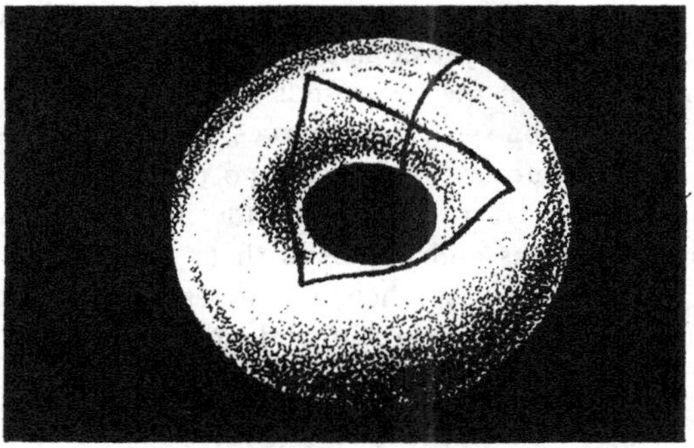

Fig. I. Torus with triangle and intersecting straightest line.

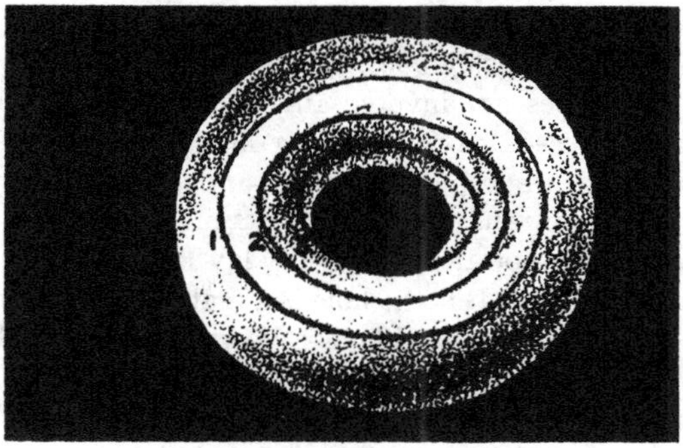

Fig. II. Non-intersecting closed curves on a torus.

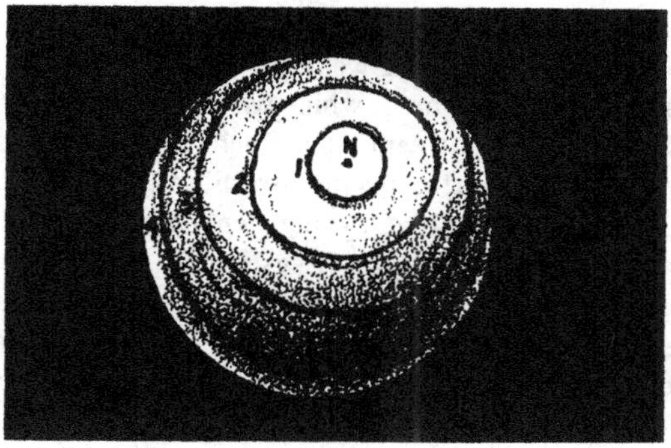

Fig. III. Relativity of the enclosure of circles on a sphere.

Plate 1.

"left" of the curve there are connecting lines which remain in the surface of the torus and yet do not intersect the curve.[1]

The ring property of the torus finds it conceptual expression in the existence of such non-separating curves; this is a holistic property preserved when the torus is uniquely and continuously mapped upon a different surface. A rubber band, for instance, has the holistic properties of the torus and preserves them even if it is twisted and distorted. Due to its other holistic properties the torus cannot be mapped upon the plane uniquely and continuously, nor upon the sphere, from which it is also topologically different.

In topology, mathematics deals with the purely qualitative properties of geometrical figures (which shows, by the way, that the statement "mathematics is a purely quantitative science" is false). Mathematics characterizes topological equivalence by the possibility of transforming uniquely and continuously one surface into the other, i.e., by a transformation that does not involve any metrical considerations. Thus mathematics succeeds in formulating analytically those geometrical properties which are typically visual and seem to defy conceptual formulation. By means of new concepts mathematics teaches us how to visualize such properties. In everyday language we call the torus a surface with a hole. But the hole is a matter of the third dimension; the *surface* of the torus has no hole. When we walk on the surface we always find ourselves in an uninterrupted environment. Nevertheless, the phenomenon we called the hole of the torus manifests itself in experiences on the surface; we formulated these experiences by the existence of curves which cannot be contracted to a point and among which obtains an undetermined betweenness relation. Even for the surface of the torus as a whole we must readjust our visualization in a way similar to the readjustment which we found necessary in connection with the relation of betweenness. Such considerations show that, indeed, "percepts without concepts are blind." This striking remark of Kant's is better illustrated by mathematical analysis than by the argumentation of his philosophical system.

Considerations of this kind will therefore be our guide in attempting to transfer topological properties to the three-dimensional domain. Since we can reconstruct the metrical properties of curved surfaces in

[1] There are cases in which a curve cannot be contracted to a point, yet it divides the surface into two separate domains, such as the ring curve on an infinite cylinder. The corresponding ring curve on the torus, however, has the same properties as the curves drawn in Fig. II of the plate.

the three-dimensional realm, we should also be able to recognize their topological properties in three-dimensional spaces. The simple connectivity properties of Euclidean space will be a special case, and we must analyze spaces with different connectivity properties. In these considerations we need no longer separate the question of physical realization from that of visualization. Since we found that the problem of visualization can be solved in connection with the construction of perceptual experiences, we shall deal with both questions at once. We shall follow Helmholtz' method; for "visualize" Helmholtz gave the definition: ". . . that we are able to imagine the series of perceptions we would have if something like it occurred in an individual case."[1] We ask therefore: What would we experience if space had different topological properties? By imagining a torus space we shall try to answer this question.

For the presentation of the physical facts to be described we shall use the same method employed successfully in the preceding section. We assume at first that the space is Euclidean and describe all observations according to the scheme G_0+F. Only later shall we go to $F = 0$ and G. We wish to construct the three-dimensional analogue to the nonseparating curves on the surface of the torus.

Fig. 8 is to be conceived three-dimensionally, the circles being cross-sections of spherical shells in the plane of the drawing. A man is climbing about on the huge spherical surface 1; by measurements with rigid rods he recognizes it as a spherical shell, i.e., he finds the geometry of the surface of a sphere. Since the third dimension is at his disposal, he goes to spherical shell 2. Does the second shell lie inside the first one, or does it enclose the first shell? He can answer this question by measuring 2. Assume that he finds 2 to be the smaller surface; he will say that 2 is situated inside of 1. He goes now to 3 and finds that 3 is as large as 1.

How is this possible? Should 3 not be smaller than 2? If the geometry were Euclidean this would be the case. But this consideration need not influence the observations. The physicist will explain his measuring results by a contraction of the rod; his rod and also his body contract so that 3 appears larger than 2. He can avail himself of this hypothesis.

[1] *Schriften zur Erkenntnistheorie*, edited by Hertz and Schlick, Berlin 1921, p. 5. This formulation by Helmholtz, in connection with his examples constructed according to this principle, has opened the way for the solution of the problem of the visualization of geometry.

He goes on to the next shell and finds that 4 is larger than 3, and thus larger than 1. His rod has further contracted; 5 he again finds to be as large as 3 and 1.

But here he makes a strange observation. He finds that in 5 everything is familiar to him; he even recognizes his own room which was built into shell 1 at a certain point. This correspondence manifests itself in every detail; he sees his own handwriting on the paper upon the table, and his teacup stands half-empty where he left it. He is quite dumbfounded since he is certain that he is separated from

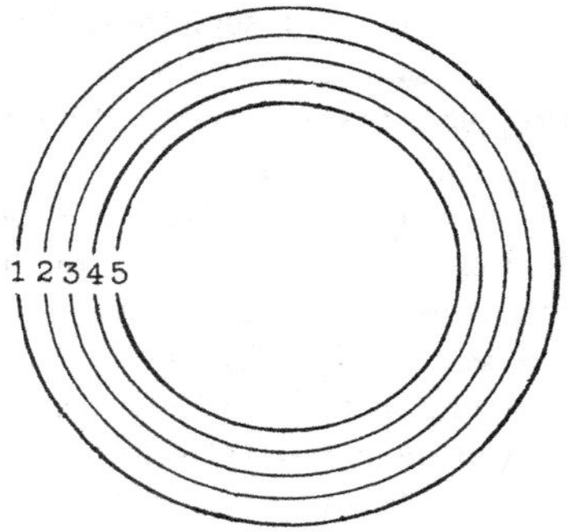

Fig. 8. Concentric spherical shells.

surface 1 by the intervening shells. He must assume that two identical worlds exist, and that every event on surface 1 happens in an identical manner on surface 5.

It suddenly occurs to him that at this moment his own double sits in surface 1 puzzling about the same things as he. In order to test this idea he makes a crucial experiment. He writes down his thoughts on a sheet of paper, adds a code word, locks the paper in a drawer, puts the key in his pocket, and leaves shell 5. He investigates it once more and finds that it is completely enclosed by shell 4. He then returns to 1, making sure all the time that every shell is situated between two other shells. Arriving at 1 he finds his room, opens the drawer with the key he put into his pocket, and recognizes on the slip of paper the same words which he had written down in shell 5.

What kind of a world will he imagine? If he retains Euclidean geometry he will have to accept the duplication of all happenings including his own person. Not only a duplication, however; he would also find that by wandering from 1 in the "outside" direction or from 5 in the "inside" direction he encounters the same things. He can never reach the center of all the shells because as he approaches it all objects as well as his own person contract continuously; the center, for him, has the property of an infinite distance. The world consists of an infinite number of periodically equivalent spherical shells. Within each interval, say from 1 to 5, everything happens according to the usual physical laws, but from there on, the same happenings recommence. To every point of one interval of shells corresponds a respective point in all the other intervals; yet there is no boundary, and all transitions are continuous.

This is a description of experiences in a torus space presented in Euclidean geometry. We notice that the addition of a universal force F to G_0 does not suffice; in addition, a *causal anomaly* occurs, consisting in the spatial periodicity of all happenings. The interdependence of all events at corresponding points cannot be interpreted as ordinary causality, because it does not require time for transference and does not spread as a continuous effect that must pass consecutively through the intermediate points. Only *within* every shell does normal causality hold; the interdependence of the shells is like some kind of preestablished harmony. It may be left open whether this preestablished harmony is to be conceived as an instantaneous coupling of distant events, i.e., as an action at a distance, spreading without intermediate effects, or whether it is to be regarded as a parallelism of events which had "by chance" the same initial conditions and since then have been running down like synchronized watches. Such a distinction is merely a difference in interpretation.[1] The state of affairs underlying the two interpretations is essentially different from the normal laws of nature, and we speak therefore of a causal anomaly. We may no longer write our formula "G_0+F" but

$$G_0+F+A$$

where A stands for the causal anomaly.

On account of this result, physics arrives at a strange situation. The principle of causality is one of its most important laws, which it will not abandon lightly; preestablished harmony, however, is incompatible with this law. In physics the transition to the geometry of the

[1] The latter interpretation is possible only from the standpoint of determinism, for otherwise the permanence of the strict parallelism would be infinitely improbable.

torus will thus be preferred. In this conception the shells 1 and 5 are *identical*, and the world does not exist in periodic sections in space, but only once, in the shape of a torus. With the transition from G_0 to G not only the universal field of force F but also the anomaly A disappears, a consequence which is a strong argument for preferring G.

The question arises whether this result is compatible with the relativity of geometry proved above. This relativity was based upon theorem θ (p. 33) which stated the possibility of mapping geometries upon each other. As we said at the beginning of the present section, a unique and continuous mapping is possible only for geometrical structures having the same topology; theorem θ is correct only within these limits. If topologically different spaces are mapped upon each other, neighborhood relations will be disturbed at some places. The mapping of the torus space presented above corresponds in two dimensions to the case in which the surface of the torus is cut open along one of the curves drawn in Fig. II (Plate 1) and shaped into a plane circular ring; under these circumstances the surface of the torus will be deformed and the points along the cut will be separated from adjacent points. The two edges of the ring correspond to each other, i.e., they are assigned identical points on the torus. The continuity is reestablished when an infinite number of smaller and smaller rings are arranged concentrically within one another; but then the mapping will cease to be unique in one direction. The violation of continuity or uniqueness corresponds in the physical interpretation to a causal anomaly, in this case to the preestablished harmony. If perfect freedom of choice of geometry is to be preserved as a *conditio sine qua non*, causal anomalies must be reckoned with occasionally.

Do we have to renounce Euclidean geometry in such a case? We do not have to, because no one can prevent us from believing in a preestablished harmony; if we admit it, Euclidean geometry is saved. We can say, however: if normal causality is to be retained, Euclidean geometry must under certain conditions be excluded for physical space. In this case an additional condition must be normalized, if the statement about the geometry of space is to have an objective meaning. It is clear that physics will require normal causality, if it introduces the normalization $F = 0$. This assumption had been implicitly made whenever the geometry of space was discussed; without this assumption all statements about the geometry of physical space would become ambiguous.

The torus space was analyzed thoroughly because such a discussion furnishes an important argument against the aprioristic philosophy of space. It was said above that the aprioristic philosopher cannot be prevented from retaining Euclidean geometry, a consequence which follows from the relativity of geometry. However, under the circumstances mentioned he faces a great difficulty. He can still retain Euclidean geometry, but he must renounce normal causality as a general principle. Yet for this philosopher causality is another *a priori* principle; he will thus be compelled to renounce one of his *a priori* principles. He cannot deny that facts of the kind we described could actually occur. We made it explicit that in such a case we would deal with perceptions which no *a priori* principle could change. Hence there are conceivable circumstances under which two *a priori* requirements postulated by philosophy would contradict each other. This is the strongest refutation of the philosophy of the *a priori*.[1]

What can be said now about the possibility of visualizing the torus space? The same considerations that were explained in the preceding section apply to the metrical deformation of the measuring rods. This deformation can be visualized by means of a readjustment to a different congruence. The identification of the spherical shells 1 and 5 in Fig. 8, however, presents a greater difficulty. It is quite certain that we would regard the individual objects on the shells as identical; when perceived they are identical in the usual sense. The problems of visualization arising in connection with the mutual enclosure of the spheres will be discussed later.

A topologically different space is the spherical space, which is particularly interesting in that it does not represent merely a possible form of physical reality like the torus space, but, according to Einstein, corresponds to real space. In order to imagine it, we again construct the visual experiences in terms of a two-dimensional analogue. However, we shall choose much smaller dimensions for our model than those of the Einsteinian space of the universe; otherwise we would not be able to describe visual experiences noticeably different from those in Euclidean space.

On a spherical surface, as on a plane, every closed curve can be contracted to a point. Still, there is a difference: the curves can be contracted *in both directions*. They can be contracted in the direction

[1] My refutation of the Kantian system, presented in *Relativitätstheorie und Erkenntnis a priori*, Berlin, 1920, p. 29, is based on this idea; the present exposition is a still better example of a contradiction between *a priori* principles than the one presented in the earlier publication, which is not quite correct.

of the north pole as well as in the direction of the south pole, because the surface of the sphere is closed. On the spherical surface it cannot be stated, therefore, which of two mutually enclosing circles is the outer one. If two circles 1 and 2 close to the north pole N (Fig. III, Plate 1) are considered, the larger circle 2 seems to be on the outside; but circle 2 may be regarded as an intermediate step in the process of contracting circle 1 to a point, if the circles 3, 4, 5 are contracted in the direction of the south pole. Enclosure is a topological concept, and size cannot be an indication of enclosure; on the sphere, therefore, we may not speak of a one-sided enclosure of circles as we would on the plane. The concept of enclosure is relative; there is only an "enclosure with respect to a given point of contraction." *The relativity of enclosure expresses conceptually the finiteness and closedness of the spherical surface.*

This idea must be transferred to three-dimensional space. We should not look for an analogue to the model in which the finite surface of the sphere is imbedded in a three-dimensional space; finiteness does not mean a restriction to an island in a larger surrounding world. If we limit ourselves to the number of dimensions of the structure itself, there exists nothing else besides it, and there is no place in this world which we cannot reach. Finiteness is rather expressed in the specific topological relations of all space points. Corresponding to an exterior and an interior curvature, exterior and interior holistic properties must be distinguished; only the interior holistic properties will be investigated here.

Spherical surfaces in three-dimensional space correspond to the circles on the surface of a sphere. Let us analyze their relations of enclosure by mapping the three-dimensional structure upon Euclidean space, i.e., by drawing the corresponding distorted figures in Euclidean space. We must proceed in this manner because we intend to draw the spherical space. "Drawing" means mapping upon a small area, and since small areas in a spherical space are nearly Euclidean, the inhabitant of the spherical space will not be able to draw pictures different from the ones we are going to present.[1] Yet we are mostly interested in discovering what he will *see.* From our drawing we shall be able to infer his perspective, which is very different from the Euclidean one. We then can draw without distortion pictures that represent his perceptions, because we merely have to draw plane

[1] "Drawing" is conceived here as a mapping upon a small three-dimensional volume; the pictures in the plane of the drawing are projections of this figure.

figures which, when projected upon the retina, will furnish the same pictures as those actually occurring in spherical space.

The mapping of the spherical space will be accomplished by a *stereographic projection.* We start with the projection of a two-dimensional spherical surface. From the center of the projection P (Fig. 9) all points of this surface will be projected by light rays upon the opposite tangential plane; the top view of the resulting figures is drawn in the bottom part of Fig. 9; its center is the point O, opposite to P,

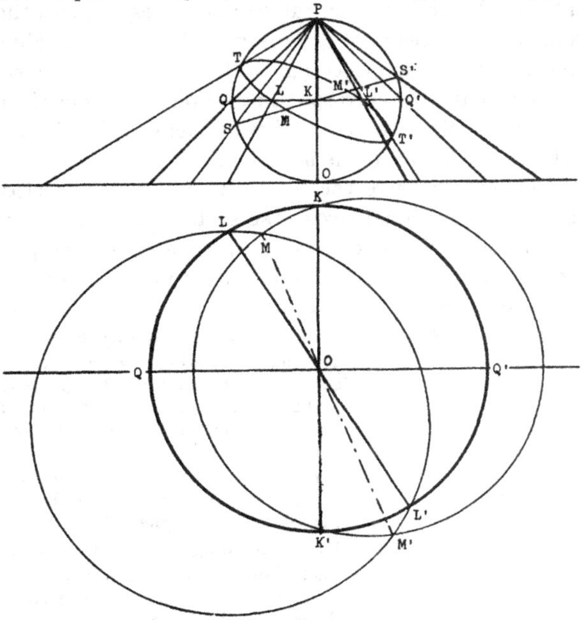

Fig. 9. Stereographic projection of the surface of a sphere. If conceived as a cross-section through spheres, the top view (the lower drawing) is the stereographic projection of spherical space.

whereas P itself is mapped upon infinity. All circles going through P become straight lines in the plane of the projection. In particular, the great circles going through P become *central straight lines*, i.e., straight lines going through O. The equator QQ' becomes the *fundamental circle* $QKQ'K'$, the center of which is O. It can be shown, however, that any circle of the sphere is transformed into a circle of the plane—a result not intuitively evident, but known to mathematicians since the time of the Greeks. Great circles, in particular, are mapped as circles intersecting the fundamental circle $QKQ'K'$ in two diametrically opposite points, because they possess this property on the sphere; we call them *main circles.* In the top view (i.e., the bottom part) of Fig. 9, two of these main circles are drawn, constructed as the projections of the great circles SS' and TT', whose front view appears in the top part of Fig. 9.

It follows from the unique correspondence given by the mapping that the system of figures of the plane formed by main circles, the fundamental circle, and the central straight lines satisfies the axioms of two-dimensional spherical geometry. This fact enables us to transfer the stereographic projection to three-dimensional space. One would assume that the corresponding three-dimensional structures satisfy the axiomatic system of three-dimensional spherical space, and in fact this can easily be demonstrated.

We need merely conceive the top part of Fig. 9 as a cross-section through a three-dimensional space; the system of spheres, represented in the drawing by circles as cross-sections, is the stereographic projection of the spherical space. In the following table we write down the coordination given by the projection:

Element of the spherical space:	*Representation in Euclidean space:*
Plane	The fundamental sphere, or plane through the center O of the fundamental sphere (= central plane), or sphere which intersects the fundamental sphere in a great circle (= main sphere).
Straight Line	Straight line through O (= central straight line), or circle of intersection between central plane and main sphere (= main circle).
Point	Point, including the infinitely distant point.
Congruent figures	Such figures as can be transformed into each other by a spherical transformation that produces a fundamental sphere of the same size.

Let us sketch briefly an elementary proof that the system of elements on the right satisfies the axioms of three-dimensional spherical geometry.

It is clear that on every central plane the same relations exist as in the plane of the top view drawing of Fig. 9, because every circle resulting from the intersection of a central plane and a main sphere must possess the properties of the main circle defined above. (The circle of intersection between the fundamental sphere and a main sphere is a great circle, according to the definition of a main sphere; it is intersected in two diametrically opposite points by the great circle resulting from the intersection of a central plane and the fundamental sphere. Since these points belong both to the central plane and the main sphere they must be situated on the circle of intersection of these two elements.)

Three more theorems are to be proved. First: The main circles of every main sphere must satisfy the two-dimensional spherical geometry. Though these main circles are not great circles of the main sphere in the Euclidean sense, they follow the same system of axioms as the great circles, because they result from the projection of the totality of great circles of the central sphere upon the corresponding main sphere. Second: It can be proved that any two main spheres intersect in a circle the plane of which passes through O. It is obvious that the straight line MM' connecting the points of the intersection of two main circles passes through O, since M and M' correspond to diametrically opposite points of the sphere indicated in the front view of Fig. 9. Since the straight line MM' falls in the plane of the intersection of the two main spheres, O falls in this plane. Third: It can be proved that two main circles that have a common point determine a main sphere to which they both belong. We first show that two main circles α and β (not drawn in Fig. 9) that have a common point G must have another common point G', which is the opposite pole to G. This is proved by the use of a third main circle γ which lies in the same central plane as α and in the same main sphere as β. We next construct a sphere through G and G' and one point each of α and β; the sphere determined by these four points must contain α and β. It must also be a main sphere because it passes through G and its opposite pole G'.

The other axioms can easily be read from the model. For instance, the nonexistence of parallels is represented by the theorem that any two main circles belonging to the same main sphere must intersect each other. There are also straight lines that are skew, i.e., which do not intersect, and yet cannot be incorporated into one plane. These are represented by two main circles linked to each other in the manner of a chain. Two planes must always intersect, because there is no parallelism for planes; the model shows this feature as a property of the main spheres.

The uniqueness of the center O, the fundamental sphere, and the central planes is only apparent. It is possible to choose a different projection for which any other given point becomes the center, while one of the former main spheres becomes the fundamental sphere; other main spheres become central planes. The translation of one projection into another is achieved by a transformation with reciprocal radii, a so-called spherical transformation. This fact justifies our above definition of congruence and likewise our mode of speech with regard to infinitely distant points of Euclidean space. The infinitely distant region of Euclidean space is properly regarded as a plane. Since, however, it is equivalent to a single point located in finite space—according to the transformations given here—we speak of an infinitely distant point.

By means of the stereographic projection it is easy to construct the relations in spherical space. Since we are looking for an analogue to the relativity of enclosing circles, Fig. III, (Plate 1) we shall consider the following structure.

Let us imagine in space two large spherical shells I and II, made of sheet metal, which enclose each other and are rigidly connected by beams. An observer climbs around between the shells; however, he cannot pass through them but is restricted to the space between the spheres. He intends to determine which shell is the outside one.

In order to visualize his experience we construct the following figure. In the stereographic projection we draw two concentric spheres I and II, the top view of which can be seen in Fig. 10. Let us assume that they are symmetrical to the fundamental circle indicated by a dotted line; i.e., in the original sphere (front view of Fig. 9) they correspond to circles of latitude lying symmetrically with respect to the equator QQ'. The observer is stationed at A on the fundamental circle. In order to make our problem precise we assume that light rays move along straightest lines in space.[1] Hence we can determine by means of main circles and main spheres what is visualized by the observer, just as in Euclidean space a representation of his perceptions is ascertained

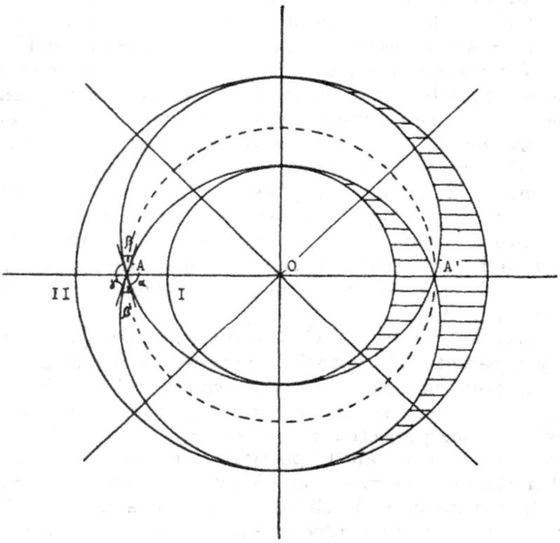

Fig. 10. Stereographic projection of spherical space: perspectives of an observer at A.

by means of straight lines and planes. We draw two main circles through A which are tangential to the two circles I and II; they furnish the angular perspective for A as do corresponding lines of projection in Euclidean space. Since in every plane through A and O the same relations hold, we may conceive Fig. 10 as a cross-section through a three-dimensional figure which results from a rotation around the axis

[1] It should be pointed out that this assumption does not correspond to the conditions of the general theory of relativity, since according to it light rays move along four-dimensional straightest lines, so that even in the static field of gravitation this phenomenon does no lead to straightest lines of three-dimensional space. Cf. Axiomatik der relativistichen Raum-Zeit-Lehre, p. 128 (hereafter referred to as A.) The deviation, however, is very small in the case of a weak curvature.

AOA'. If the view of the observer is confined to the angular space α he will see shell I; in the angular space γ he will see shell II, and in the angular space β he will see the empty space between the shells. The shaded angular space on the right side of the figure is invisible to him, because it is hidden by the shells; it is the "shadow area" for A.

For the purpose of constructing the picture of his perceptions we must follow the cone of light rays which begins at A; since the stereo-

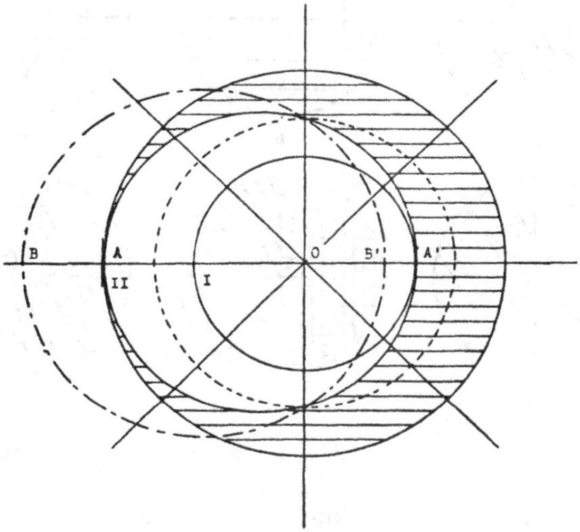

Fig. 11. Stereographic projection of spherical space: perspectives of an observer at A.

graphic projection reproduces the original angles, this cone is immediately given by the tangents at A. It is a double cone symmetrical to the line AO. His perceptions are obtained by the intersection of this double cone with a plane of projection, which must always be assumed to lie perpendicular to the direction of view. Fig. 12 shows the views which are seen in the three directions: (a) from A towards I, i.e., along the central axis of α; (b) from A along a direction perpendicular to the first, i.e., along the central axis of the adjacent angular space β; (c) from A towards II, i.e., along the central axis of γ. In these figures the shells are distinguished from each other by different shadings; Fig. 12*a* shows the shading of shell I, Fig. 12*c* that of shell II.

In Fig. 11 we have drawn the perspective relations for an observer who stands next to a shell, at A. The double cone of the received light rays has degenerated into a plane; on one side of the plane, rays from the whole surface of shell I arrive, while the other side is completely hidden by the adjacent part of shell II. His perceptions are drawn in

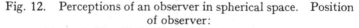

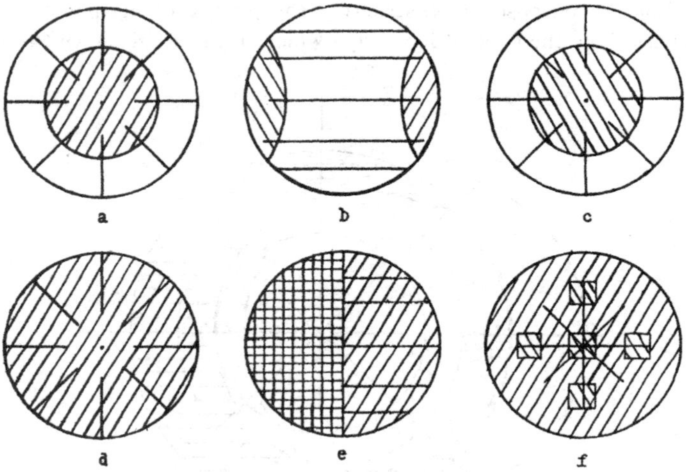

Fig. 12. Perceptions of an observer in spherical space. Position of observer:

(a) at A, Fig. 10, looking along the axis of α.
(b) at A, Fig. 10, looking along the axis of β.
(c) at A, Fig. 10, looking along the axis of γ.
(d) at A, Fig. 11, looking towards I.
(e) at A, Fig. 11, looking perpendicular to AO.
(f) at O, Fig. 11, looking in any direction.

Fig. 12*d* and 12*e*; 12*d* corresponds to the central direction towards I, 12*e* to the direction perpendicular to that of 12*d*.

We can now imagine the visual experiences of the observer. From the space between the shells he sees both spheres as *convex* surfaces; i.e., by looking towards the spheres he discovers that light rays do not glide along the surface, and that within the space between the shells there is no connecting light path between two points of the same surface. If he stands in the middle of the shell space looking towards I, he sees in front of him the convex hemisphere of this surface surrounded by free space; when he turns around, he sees shell II in the same manner, i.e., its convex hemisphere surrounded by free space. The two diagrams 12*a* and 12*c* represent these perceptions, and we may conceive them three-dimensionally as in Fig. III (Plate 1). Representations of the supporting beams must be added; they intersperse the shells vertically and are indicated in Fig. 12*a*—12*f*. When he turns his head, the beams seem to rotate with respect to each other just as in the usual Euclidean space the straight-line junctions of the walls and ceiling of a long room seem to rotate with respect to each other as one turns one's head from a direct overhead view to the perspective of the far end of the room. From this point the total perspective does not show the spheres as enclosing each other, but as situated side by side. However, when the observer changes his position, the surfaces are bent open in a strange way: with increasing distance from a shell it flattens out, and the observer sees larger and larger areas, i.e., more than its hemisphere. If he stands immediately beside a shell, he sees the entire surface of the other shell and it looks as flat as a plane. There is no empty space to be seen from this position of the observer, and the beams lie in front of the shell. This is the view given in Figs. 12*d* and 12*e*; in 12*e*, which corresponds to the perpendicular direction, the left half of the field of vision is hidden by the small part of the convex shell II against which the eye of the observer is pressed.

The empty space between the shells shows remarkable features. Since all light rays originating from the observer are gathered in A' (Fig. 10), a real optical image of the observer occurs in this spot. It is true that every direction of view leads to A', but this holds strictly for one point only; points adjacent to A' are seen in one direction only. The environment of A' is then coordinated to the total bundle of rays originating from A. Because of the natural limitation of the angle of vision inherent in the structure of the eye, only that section will be perceived which is formed by radial lines from the back of the head of the observer: the observer thus sees the picture of the back of his own head in a huge distortion all over the space, filling the entire background. In Fig. 12*a–c* we must imagine the nonshaded area filled with an enlarged picture of one's own hair.

Let us add the following remarks about the accommodation of the eye with respect to spatial depth. The optical image results normally from the fact that a bundle of *divergent* light rays originating from the object hits the lens of the eye and is there changed into a *convergent* bundle, the point of convergence of which, or point of the image, lies beyond the focus. In spherical space, a bundle of light rays emanating from an object-point at a distance greater than a quadrant *converges* upon the eye. The lens gathers such rays into a point lying between the lens and the focus. The pattern of the rays is the converse of the one resulting for a magnifying glass. Consequently, the lens of the eye, when looking at such distant points, must be flattened to a higher degree than corresponds, for the normal case, to the adjustment to infinity. The normal eye will therefore be near-sighted in spherical space and require the correction of concave glasses. In binocular vision the two eyes must diverge for the same reason. This means that for depth perception the same criteria exist as in ordinary space, however in a more pronounced fashion, and the judgment of depth is therefore possible in the same way as in normal vision. The image of the back of the head of the observer is distinctly visible as the most distant object; this impression is confirmed if closer objects partially hide the back of the head.

Measurements by means of rigid rods would confirm the symmetry of the whole picture. An observer would find the two spherical shells equal in size, and the connecting beams equal in length.

Let us go a step further and analyze the perceptions in the interior of the shells. The observer might discover a window and look through into the shell. He sees very clearly the interior of the shell; the beams run crosswise from the wall toward the center where they intersect. This aspect only confirms the symmetry, because on opening a window in the other shell he observes the same phenomena: he sees the interior of the shell and the beams meeting in the center. The greatest surprise, however, awaits him when he climbs into one of the shells. Let us assume that he has discovered additional windows in the shell and has opened them; if he stands now in the center of the shell, i.e., at the point of intersection of the beams, he sees not only the shell all around him but also, through the open windows, corresponding wall sections of the other shell at the same distance in every direction, so that he finds himself at the same time in the center of the second shell. This perception is represented in Fig. 12*f*; for every direction of view the picture is the same. Fig. 11 furnishes the explanation for what is seen by an observer standing at O. The shaded shell of Fig. 12*f* is supposed to be concave; here the visual impression of the concentric position of the shells corresponds precisely to the respective perceptions in Euclidean space. If the observer occupies an eccentric position (at B′, Fig. 11) he also sees himself in the interior of both shells, i.e., from this position too, he looks in every direction, first at shell I and then through the windows at shell II behind it. The picture of the second shell, however, is strangely distorted, since, as is evident from the study of the main circle through B and B' in Fig. 11, the larger calotte of shell II appears behind the smaller calotte of shell I, and vice versa. In the interior of shell II the corresponding pictures would be obtained in inverse order; an observer at B has the same perceptions as an observer at B' (Fig. 11), only the order of the shells is changed. The perceptions corresponding to Fig. 12*f*, where the shading would have to be exchanged, belong to an observer whose position in the stereographic projection (Fig. 11) lies at infinity.

The observer has the following visual experience of the relativity of the mutual enclosure of the shells: At one time he sees shell I inside shell II; at another time he sees shell II inside shell I; besides, there are intermediate positions from which he sees the two shells not as concentric but as side by side with separate centers.

So far we have proceeded by developing from the properties of spherical space the perceptions of an observer stationed in such a space. We shall now use the opposite method. Assume that an observer has the perceptions described above; what would he infer? Only in this form is the problem accessible to an epistemological analysis. So long as we start from a certain state of the universe and infer perceptions from it, an aprioristic philosophy may contend for any number of reasons that such a state of the universe does not exist. As soon as we start from perceptions, however, the objection disappears, because nothing can be prescribed for perceptions. No *a priori* postulate can exclude the possibility that some person may at some time have certain perceptions. Only the *interpretation* of such perceptions is controversial. The interpretation concerns the inference from the perception to the physical world; this inference is not unique, because there are different geometrical interpretations for the same perceptual data. Let us study the two most important interpretations, those of Euclidean space and of non-Euclidean spherical space.

The interpretation is simple for non-Euclidean geometry. In this interpretation we deal with spherical space; there is no absolute "exterior" for the spheres; each of them is the exterior one with respect to the corresponding point of contraction. The two points of contraction are given in our model by the centers of the scaffolds. The space is finite, but nowhere has the character of an island, i.e., every point can be reached. Even a visual image is possible; for each of the two spherical shells we may employ the old concepts of inside and outside and may imagine them as curved in the conventional sense. It is true that the visual images change and that there does not exist *one* visual image reproducing the entire space, such as exists for the two-dimensional spherical surface that can be perceived in its entirety at a glance. However, such a plurality of images occurs merely because we ourselves are the observers, standing in the interior of the space and thus may not expect a visual image which necessarily can result only from being imbedded in a higher dimensional manifold. One visual image comprehending the entire space is not even possible in Euclidean three-dimensional space. It is quite possible, however, to attain a general impression of spherical space by *visual integration*, so to speak, i.e., by looking around and pacing it off. The visual adjustment is achieved when the resulting changes in the perceptions—the above described transition from a convex surface by way of the plane, to a concave surface—are no longer experienced as a folding inside out of the surface but as a change in perspective.

For the interpretation in Euclidean geometry we need more complicated images in order to account for the perceptions depicted above. In Euclidean geometry only one side of each shell can be considered a closed space; on the other side is the surrounding exterior space. The two shells can be conceived as being concentric or as lying side by side; in the first case the exterior space would correspond to the interior of a sphere in the non-Euclidean conception; in the second case, to the space between the shells. Universal forces have to be assumed for this interpretation: metrical deformations of rods and light rays cause deviations of our perceptions from the foregoing description. It is a representation of the objective world which treats the manner in which we draw spherical space in Figs. 9–11 as binding for space itself: the space is actually Euclidean, but the measuring instruments change their size when transported, and light rays travel along curved paths so that deviating perceptions result. Whereas such a conception is natural for the drawing—here the relations of congruence of rigid rods and the paths of light rays for small dimensions are unlike the corresponding relations for large dimensions—it assumes a peculiarly empty character for large dimensions. The difference in the relations of congruence cannot be interpreted as the result of differential forces upon the instruments of measurement, but must be explained as a universal deformation, which affects all things without exception and cannot be measured as a deviation from an actual physical state.

The essential difficulty of the Euclidean interpretation lies in the fact that the infinity of this space is physically attainable. This interpretation need not necessarily localize infinity at a marked point such as the point of intersection of the beams; since this point is too obviously realized, the attempt will be made to assume infinity in another domain of this space. The physicist would proceed by carrying a beam of finite thickness through the entire space; the interior of this beam would have to reach every point of the space at least once. Euclidean geometry will then have to admit that a material structure of previously finite dimensions can be laid down in such a way that it is situated at the same time in a finite and an infinite region without breaking. It must assume, furthermore, that a body can run through infinite Euclidean space in a finite time and return from the other side to the point of its departure. This conception contains causal anomalies; we are confronted with the same case G_0+F+A which we described above and which, if admitted, complicates the causal relations unnecessarily.

For physics there is a definite distinction between finiteness and infinity; whereas the mathematician speaks of the infinitely distant plane or the infinitely distant circle and manipulates them just like finite structures, the physicist with his real measuring rods is bound to finiteness. He cannot coordinate physical things to concepts of infinite structures; rather, by translating statements about infinite structures into statements about finite structures, he must find out whether they possess physical meaning under these conditions. Infinity of space to him means that there are no limits for the laying down of measuring rods and that by laying down measuring rods on a straight line the point of departure will not be reached after a finite number of operations. But it is a violation of all previous notions of causality underlying all our physical statements to say that a body runs through infinity, or that a body located at infinity causally affects bodies located at some finite point. There may be cases that would compel us to accept such causal anomalies; but so long as it is possible to exclude anomalies we shall retain normal causality in the sense of a postulate. Similarly, the physicist cannot admit transformations that map infinity upon a finite domain; for him, only one of the two geometries related by such a transformation is admissible; which one it is, is decided by experience. In mathematics the concept of topology is ordinarily used in a wider sense than was defined above; in the transformation group based on topological equivalence, certain singularities are admitted so that topologically the difference between finitude and infinity disappears.[1] For physics, however, the narrower concept of topology must be carried through, strictly limited to

[1] Cf. F. Klein, *Elementarmathematik vom höheren Standpunkte aus*, Berlin 1925, Vol. II, p. 142–143.

transformations that are everywhere continuous and unique in both directions.

Another objection seems to result from the mathematical fact that the anomaly occurs strictly at only one point, as illustrated by the stereographic projection of the sphere described above; and what occurs at a mathematical point cannot be determined by the physicist because of the limits of exactness of measurement. If in the space considered the infinite point has not been found, one can suppose it to be within an area which appears small to us because of the increased size of all physical objects, and which we did not occupy with matter while carrying around our beam; this point may, for instance, be in the pores of a wooden beam of the scaffold. This contention neglects the fact that in physics results may be interpreted in a way different from those in mathematics. Physics may replace the inexactness of measurement by a probability inference: if, under the assumption of a spherical space, I cannot find an anomaly, whatever experiments I perform, I assume with probability that it does not exist. Probability statements are indispensable once we agree to base our decision about the geometry upon measurements. The assertion of a spatial singularity has physical meaning only if it is in principle confirmable by inductive methods, a requirement which holds just as well for assertions about the infinity of space. The statement that physical space has the topological properties of spherical space is just as physically meaningful as a statement about metrical properties of space. *Topology is an empirical matter as soon as we introduce the requirement that no causal relations must be violated*; whether there occur causal anomalies can be decided by the usual inductive methods of physics. An example of how clearly such anomalies would be recognized consists in the experiences we would have if the surface of the earth were "redefined" as a plane. If such a "definition" were topologically admissible, there would be some point on the surface of the earth which we could not cross. It is physically meaningful, however, to say that there is no such point, although not every point on the surface of the earth has been reached by man.

Our considerations have shown that the determination of the topological properties of space are closely related to the problem of causality; *we assume a topology of space that leads to normal causal laws.* Only in this way does the question about the topology of space constitute a well-determined question. It must be called an empirical fact that there is *one* kind of topology that leads to normal causality;

and it is of course an empirical fact *which* topology yields this result. Later we shall discuss a still closer connection between space and causality. (§ 27, § 42, § 44.)

§ 13. PURE VISUALIZATION

We began the investigation of this chapter with a presentation of the mathematical development of geometry. This development showed that the problem of space bifurcates into a mathematical and a physical problem. Inquiring into the problem of the validity of the axioms, we turned to the physical problem of space. The mathematical problem of the nature of space was solved insofar as it was demonstrated that axiomatic systems contradictory to each other are equally acceptable, since mathematics concerns itself with the logical relations within these systems and not with the truth of the axioms themselves. We mentioned, however, in § 1 that this conception requires further analysis. We can now undertake this analysis, which concerns the problem of visualization of geometry. Our study of the physical problem of space has already led us to questions of this kind.

Euclidean geometry can easily be visualized; this is the argument adduced for the unique position of Euclidean geometry in mathematics. It has been argued that mathematics is not only a science of implications, but that it has to establish a preference for one particular axiomatic system. Whereas physics bases this choice on observation and experimentation, i.e., on applicability to reality, mathematics bases it on visualization, the analogue to perception in a theoretical science. Accordingly, mathematicians may work with the non-Euclidean geometries, but in contrast to Euclidean geometry, which is said to be "intuitively understood," these systems consist of nothing but "logical relations"[1] or "artificial manifolds".[2] They belong to the field of analytical geometry, the study of manifolds and equations between variables, but not to geometry in the real sense which has a visual significance.

In our previous investigations, we have examined the question of visualization and offered a visual illustration of non-Euclidean geometry, even for topologically different spaces. However, we did not state clearly whether we were concerned with physical or mathematical visualization. Since we have discussed so far only the physical

[1] H. Driesch, *Relativitätstheorie und Philosophie.* Karlsruhe 1924, p. 43–45.
[2] J. v. Kries, *Logik.* Tübingen 1916, p. 705.

problem of space, it may be argued that both our analysis of visualization based on the behavior of rigid bodies and our illustration of non-Euclidean geometry in terms of possible experiences deal only with physical visualization. And it may be maintained that there exists something like a mathematical visualization, which is not covered by our considerations. This question requires a further investigation.

It is true that in order to make possible a visualization of non-Euclidean geometry we started with the behavior of real objects and constructed imagined experiences, which led us to pictorial representation of non-Euclidean relations. However, in choosing this path, we followed the road which human visualization has taken throughout its natural development. In the behavior of rigid bodies and light rays, nature has presented us with a type of manifold which approximates Euclidean laws so closely that the visualization of Euclidean space was exclusively cultivated. There can be no serious doubt that we are here concerned with the developmental adaptation of a psychological capacity to the environment, and that a corresponding development would have led to non-Euclidean visualization, had the human race been transplanted into a non-Euclidean environment. Pedagogically speaking, the best means to accomplish a visualization of non-Euclidean geometry is therefore to picture a non-Euclidean environment. Though we see at first only changes of bodies in Euclidean space, this experience is gradually transformed, as was shown, into a genuine visualization of non-Euclidean space, in which bodies no longer change.

Does this analysis disprove the existence of a special type of mathematical visualization? Certainly not without further consideration. Reference to a biological habit does not supply an epistemological argument. We must ask what are the actual laws of the human mind, independent of their historical development. One should not forget, however, that the formulation of spatial visualization as a developmental adaptation is itself already based on an epistemological assertion, which it merely tends to emphasize, namely, the assertion that there exists a real space independent of those spaces represented by mathematics, that it is a scientifically meaningful question to ask which of the mathematically possible types of spaces corresponds to physical space, and that the "harmony"[1] of nature and reason does

[1] Kant, *Critique of Judgment*, Introduction, Chapter V.

not depend on an inner priority of Euclidean space, but that, on the contrary, the priority depends on this "harmony." Arguments that present Euclidean space as "reasonable" or "given by nature" must not be employed to establish a preference for a certain kind of mathematical space. They may be used in favor of the choice of Euclidean space for physics, in which case we might add that they also speak in favor of the opposite choice, since physical space is non-Euclidean according to Einstein. The visual preference for Euclidean space therefore cannot depend on its special suitability for the visualization of natural objects, but rather on an *inherent property* that has no connection with the outside world.

To avoid this indefinite concept of inherent property, a justification of this preference was attempted on logical grounds based on the *simplicity* of Euclidean geometry. Euclidean space has indeed certain logical advantages. Logically speaking it is simpler than non-Euclidean spaces. This simplicity is not very important, however; it is simpler in the same sense as for instance a circle is simpler than an ellipse. To claim that an ellipse is a mathematically "unreasonable" figure, inferior to a circle which belongs to a higher realm of mathematical reality, would mean a return to the mathematics of the Pythagoreans, who used arguments much closer to religious aesthetics than to mathematical science. The simplicity of Euclidean geometry is irrelevant for the philosophical problem of geometry, not only within physics (compare § 8), but also within mathematics.[1]

A logical priority of Euclidean geometry is therefore not justifiable in the sense of an epistemological superiority, and we can only base this supposed priority on a special type of geometrical visualization that has nothing to do with the perception of physical things. It is in this sense that Kant invented the concept of *pure visualization*[2] as opposed to *empirical visualization*. He too, however, was fully aware of the fact that pure visualization must be related in some fashion to empirical visualization. It is in his terminology the *form of empirical visualization*, and therefore pure visualization without any reference to reality

[1] The simplicity of Euclidean geometry is also expressed by the fact that the differential element of non-Euclidean spaces is Euclidean. This fact, however, is analogous to the relations between a straight line and a curve, and cannot lead to an epistemological priority of Euclidean geometry, in contrast to the views of certain authors.

[2] The German term *Anschauung* is translated here as *visualization*. The latter term appears preferable to the usual translation by the term *intuition*, which has a mystical connotation not intended by Kant.

is an empty notion with no epistemological significance. Since we have shown that empirical visualization can be changed to conform to non-Euclidean geometry, the same must be possible for pure visualization.

We must indeed consider our previous considerations to apply to pure visualization. Our analysis of the visualization of Euclidean geometry in § 9 holds for pure visualization as well as for empirical visualization. The investigation was concerned with abstractions and did not refer to rigid measuring rods. What else could really be meant by geometrical visualization but those structures of the imagination that appear when we think, for instance, about the diagonals of a pentagon or the shape of a closed curve on a torus? If we thought in § 11 that the non-Euclidean congruence was realized by means of material measuring rods, this too was thinking in pictures of the imagination, since the measuring rods were never actually produced. The fiction of handling material things merely facilitated visualization. We have therefore constructed pictures in the frame of pure visualization just as we draw Euclidean triangles on a blackboard. The drawing of geometrical figures is indeed nothing but the realization of geometrical figures through material things, which we have continually used in our previous discussions. Small particles of chalk are piled on a wooden board to form a triangle—what else is this but physical geometry? Every instructor who illustrates the Euclidean laws of congruence by drawing figures on a blackboard, or even by cutting out paper triangles, achieves "pure visualization" through empirical visualization. He can do this because pure visualization is nothing but a sensible quality which is realized in sense perception. This consideration explains Kant's terminology of "form of visualization." It expresses the same kind of blending of subjectivity and objectivity that occurs in color vision, which is subject to a multidimensional ordering and which can be experienced or reproduced only in sense perception. Visual forms are not perceived differently from colors or brightness. They are sense qualities, and the visual character of geometry consists in these sense qualities.

Objections have been raised against this idea on the ground that immediate sense impressions are not what is to be understood by pure visualization. The sense impression of two rails is not that of parallelism, whereas we do recognize the rails as parallel in pure visualization. One should therefore distinguish between perceptual space and the space of pure visualization. This objection, raised by

Driesch,[1] is not tenable. The fact that the two rails do not appear parallel, although they are parallel lines in an objective sense, proves nothing against the perceptual space. Rather we must ask whether there are any parallels at all in perceptual space. The answer to this question has been given long ago by psychologists.[2] There are indeed parallels in perceptual space, but their form in an objective physical space is that of two slightly curved diverging lines. We do not find a correspondence between objective and subjective parallelism, but rather a coordination of diverging lines in the objective space to parallel lines in the perceptual space, and of parallel lines in the objective space to converging lines in the perceptual space. All this is of course completely irrelevant for the problem of visualization. There do exist parallel lines in perceptual space and this fact alone is essential since it eliminates the need for a distinction between perceptual space and visualization space. Indeed, if we call the two rails parallel, this can mean only two things. First, they satisfy certain physical conditions; when we measure the distance between them with a rigid measuring rod the result is everywhere the same. Second, if this physical property is to be visualized, its representation has to be similar to the phenomenon of parallelism in perceptual space. It does not matter that the immediate sense impressions of rails do not present this phenomenon. Our assertion is not concerned with the perceptual image of rails but is a statement about their objective relation to each other, clarified by means of a visual picture. We therefore claim that the objective state of the rails does not correspond to the impression of convergence as seen in our perceptual image, but to the impression of parallelism that we might obtain from certain objectively diverging lines.

We are frequently faced with the necessity of looking for the picture required for the visualization of an object, not in the perception of this particular object, but in a different perceptual image. When we see the lines of an illuminated advertising sign from a great distance, they appear to be completely continuous. In spite of that we know that this is not the "correct" visualization of the lines. We should rather imagine the picture of a string of pearls, which describe how the lights

[1] *Ibid.*, p. 44.

[2] See F. Hillebrand, "Theorie der scheinbaren Grösse bei binokularem Sehen," *Wiener Akademieberichte* 1902, math.-naturwiss. Klasse; W. Blumenfeld, "Untersuchungen über die scheinbare Grösse im Sehraume," *Ztschr. f. Psychologie* 65, 1912, p. 252.

would appear when viewed from nearby. Or consider optical illusions. Here too, the perceived visual image is not that which is to be coordinated to the objective situation. The phenomenon of the convergence of parallels is nothing but an optical illusion. There can be no doubt that in optical illusions we do experience the visualization of the wrong picture. It is not "wrong in itself," but we can assert a discrepancy between the perceived picture and the objective state. This discrepancy again proves absolutely nothing against the fact that all visualizations are merely sense qualities of the perceptual space. Where else could we find the origin of specific visualizations? No visualization is needed for the objective assertion that the two rails are parallel. It is sufficient to state the results of measurements with rigid rods. Of course, perceptions are needed to accomplish these measurements, but for imaginary measurements they can be replaced by visual images. In these visual images, however, which can be conceived as "close-ups" of the measuring rod and small parts of the rails, the *visual* parallelism of the rails does not appear, although all of them together justify the assertion of objective parallelism. If the parallelism is also to be visualized, we must supplement our assertion by the description of certain qualities with which we are familiar from perceptual space.

Though we have used previously the concept of perceptual space, we cannot fail to mention that this term, though frequently employed, is rather unfortunate. Perceptual space is not a special space in addition to physical space, but physical space which we endow with a special subjective metric. (Each sense has a different metric.) Once this fact is realized, our argument is easily understood. That perceptual space and physical space ought to be distinguished implies that physically equal distances in physical space are not always experienced as such. There is no third kind of space, the space of visualization, because apart from the definition of congruence in physics and that based on perception there is no third one derived from pure visualization. Any such third definition is nothing but the definition of physical congruence to which our normative function has adjusted the subjective experience of congruence.

Finally, our example shows how much the visual experience of parallelism is determined by logical considerations and hence is not an absolute datum. The experiments of Blumenfeld indicated that the same lines, produced by a series of lamps in a dark room, were perceived sometimes as parallel and sometimes as divergent, depending upon the feature to which the observer attended. When the lamps were adjusted under the directive of the "directional condition" of parallelism, objective curves resulted different from those obtained when the lamps were adjusted according to the "distance condition."[1] It is exactly this variability of sense experience which was employed in our visualization of non-Euclidean geometry. Whereas the subjects in the psychological experiments mentioned were mostly passive, and restricted their activities to self-observation, the visualization of non-Euclidean geometry depends on an active concentration on the visual experience. There can be no doubt that this active participation leads to a wider range of possible variations of visualization.

The investigations of § 11, which led to the visualization of non-Euclidean geometry through an adjustment in the perception of congruences, are therefore applicable to mathematics in the same fashion as to physics. Although we have simplified the adjustment psychologically by connecting it with the idea of measuring rods of varying properties of congruence, this is by no means necessary. We could have worked directly with visual qualities by replacing the idea of the transport of measuring rods by the visual directive: "These distances which I see should be considered congruent." For abstract mathematics this procedure is equivalent to a physical coordinative definition.

This equivalence has been obscured by a certain mathematical complication which may create the impression that special conditions underlie the space of mathematical visualization and that a change in the definition of congruence will not produce a corresponding change in the visualized laws. In contrast to the physicist, the mathematician does not use the visual directive "this distance," since it would never lead him to precise visualizations. Estimates by sight are too inaccurate in an ideal geometry to take over the function of the transportation of measuring rods as used in practical geometry. Rather, the mathematician uses an *indirect definition* of congruence, making use of the logical fact that the axiom of the parallels together with an additional condition can replace the definition of congruence. He can therefore avoid the otherwise necessary reference to visual distances in his definition of congruence. Instead he introduces other basic visual elements which can be visualized more easily and which also lead to a determination of congruence. These other elements

[1] Blumenfeld, *op. cit.*, pp. 323 and 346. "Directional condition" means: one should concentrate on the apparently equal direction of the lines. "Distance condition" means: one should concentrate on the apparent equal perpendicular distances. Some subjects in the experiments showed a third variation when they concentrated on the "perpendicularity to the frontal plane."

are the analogue to the physical components of the coordinative definition.

It can easily be seen that the parallel axiom indeed presents a determination of congruence. In Fig. 13, let AB be parallel to $A'B'$ and AA' parallel to BB'. We can then define: AB is congruent to $A'B'$. With the same definition we obtain from the dotted parallel lines: $A'B'$ congruent to BC, and also AB congruent to BC congruent to CD, etc., since the concept of congruence is transitive. The Euclidean axiom of parallelism therefore determines the congruence along any straight line. This is the reason why the Riemanian generalization of geometry, through the concept of congruence, leads to the same type of geometry as the generalization introduced by Bolyai-Lobatschewsky through a change in the parallel axiom. Of course, the parallel axiom alone is not sufficient to define the comparison of differently directed line segments. As long as we restrict ourselves to the parallel axiom, it is impossible to decide whether AB is equal or unequal to AA'. For such a comparison we must introduce the *right angle*, which is the previously mentioned additional condition; i.e., a directive must be given which decides whether the four angles resulting from two straight intersecting lines are equal or not. Now we can construct rectangles, and their diagonals furnish the definition for the equality of differently directed line segments.

The visual elements introduced in connection with the definition of congruence in place of a direct comparison of length are therefore parallelism and the right angle. These are the perceptual elements of the coordinative definition. Whenever a mathematician forms visual representations, he uses coordinative definitions just like a physicist, with the difference that his coordinated objects are not physical objects but visual qualities. The fact that we are dealing here with a coordination is illustrated by the following considerations. Our definition of congruence in terms of these elements is expressed by the obvious fact that we can dispense with ruler and compass and compare line segments on a drawing board with straightedge and right triangle alone.[1] It is exactly this type of definition of congruence which we have in mind when we speak of pure visualization of Euclidean geometry, thus avoiding a direct comparison of length. The fact, however, that this procedure can also be represented by means of

[1] We have in mind the commonly used drafting device with a straightedge pivoted on a crossbar. Fixed in place by a screw, it can draw parallels in any arbitrary direction.

physical objects, namely a straightedge and right triangle, characterizes it as a coordination. We coordinate to the concepts of parallel and of right angle visual qualities which we know from the perceptual experiences of the physical objects, straightedge and right triangle. The coordinative definition required for congruence is here established in a different fashion. We use an indirect method. There is of course no objection to readjusting our visualization. Just as we adjusted our perception of congruence in the previous example, we can similarly

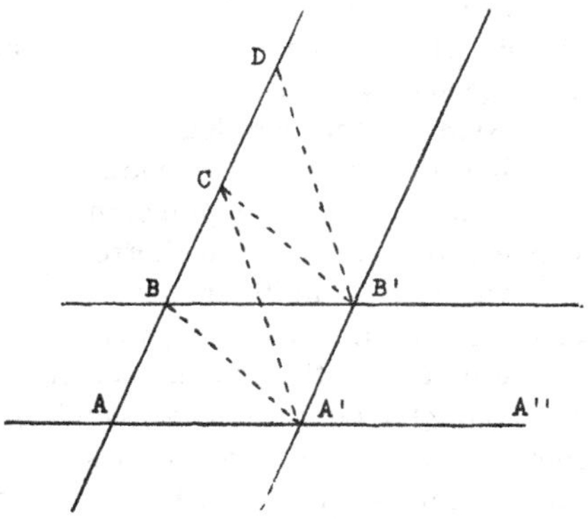

Fig. 13. Definition of congruence by means of parallels.

adjust our perception of parallelism and rectangularity. For instance, we could consider the lines in Fig. 13 as intersecting at right angles which would make angle $AA'B'$ equal to angle $A''A'B'$. The visual picture of Fig. 13 is then an admissible and noncontradictory representation of a rectangular network, with the only shortcoming that it does not fit the behavior of the straightedge and right triangle. The visual elements of a mathematically conceived space can therefore also be adjusted; and if we feel an inner resistance against such a change, this too is due to an adaptation to experiences with rigid bodies.

We have to mention another reason for the difficulties arising in connection with this adjustment. The concept of congruence in Euclidean geometry is not exactly the same as that in non-Euclidean geometry. Each geometrical concept contains implicitly all the geometrical axioms. This fact will become clearer in the next section, where we shall explain the nature of an implicit definition.

Accordingly, the content of a geometrical concept is determined only by the totality of the axiomatic system, and thus the concepts of Euclidean congruence and of non-Euclidean congruence are not identical, but play equivalent roles in the axiomatic systems. In connection with our previous remarks regarding the relation between congruence and parallelism, we can illustrate this point as follows: "Congruent" means in Euclidean geometry the same as "determining parallelism," a meaning which it does not have in non-Euclidean geometry. What is actually meant by equality of length? First it asserts a one-to-one correspondence between all points of the segments; this concept determines the higher class. The corresponding specific difference states in Euclidean geometry: ". . . such that the resulting metric corresponds to the axiom of the parallels." We can formulate this more precisely: Euclidean equality is a unique coordination of line segments, such that a curve which is equidistant from a straight line is also a shortest line and such that two shortest lines which are not equidistant will intersect at a finite distance. The first condition distinguishes Euclidean geometry from geometries with positive curvature, and the second from those with negative curvature. It is thus clear that the characterization of non-Euclidean geometry must be different. Common to the two geometries is only the general property of one-to-one correspondence, and the rule that this correspondence determines straight lines as shortest lines as well as their relations of intersection. Strictly speaking, we cannot say that the equality defined by the drawing of Fig. 6 (page 51) corresponds to the ordinary concept of equality. It corresponds only to the equivalent concept of a non-Euclidean geometry. This consideration expresses what we have called "presupposing tacit conditions." If we feel a resistance to accepting the congruence defined by Fig. 6 as an equality we react in this manner, because we miss in this equality the element which "determines parallelism." This goal, however, we cannot achieve, because it is logically impossible, and therefore must be renounced in any attempt to adjust our visualization to a non-Euclidean congruence.

We can summarize our results as follows:

There is no pure visualization in the sense of the *a priori* philosophies; every visualization is determined by previous sense perceptions, and any separation into perceptual space and space of visualization is not permissible, since the specifically visual elements of the imagination are derived from perceptual space. What led to the mistaken conception of pure visualization was rather an improper interpretation of the *normative function*, which we have recognized in § 9 as an essential element in all visual representations. Indeed, all arguments which have been introduced for the distinction of perceptual space and space of visualization are based on just this normative component of the imagination. Although it is usually granted that the visual picture of two parallels on the blackboard is not different from the picture of two lines, the prolongation of which would intersect in the sun, the visual insight into the Euclidean axiom of the parallels is generally considered to be very compelling. And rightly so: in spite of all the limitations of visualization, we can visualize the Euclidean axioms rather well. Consider, for instance, the axiom that the straight line is the shortest connection between two points. It is intuitively certain that a straight line is shorter than any line however slightly curved; this insight is due to the peculiarity of human thinking that it can draw strict conclusions from vague visual pictures; this is an important ability of the human mind which we put to continual use. It is therefore impossible to disprove the existence of pure visualization on the basis of a lack of clarity in visual pictures. On the contrary, this argument has given a new impetus to the thesis of pure visualization, since it has been interpreted in terms of a distinction between the vague perceptual space and the precise space of visualization. The main objection to the theory of pure visualization is our thesis that the non-Euclidean axioms can be visualized just as rigorously if we adjust the definition of congruence. This thesis is based on the discovery that *the normative function of visualization is not of visual but of logical origin* and that the intuitive acceptance of certain axioms is based on conditions from which they *follow logically*, and which have previously been smuggled into the images. The axiom that the straight line is the shortest distance is highly intuitive only because we have adapted the concept of straightness to the system of Euclidean concepts.[1] It is therefore necessary merely to change these conditions to gain a correspondingly intuitive and clear insight into different sets of axioms; this recognition strikes at the root of the intuitive priority of Euclidean geometry. Our solution of the problem is a denial of pure visualization, inasmuch as it denies to visualization a special extra-logical compulsion and points out the purely logical and nonintuitive origin of the normative function. Since it asserts, however, the possibility of a visual representation of all geometries, it could be

[1] Compare this with p. 100.

understood as an extension of pure visualization to all geometries. In that case the predicate "pure" is but an empty addition, since it denotes only the difference between experienced and imagined pictures, and we shall therefore discard the term "pure visualization." Instead we shall speak of the normative function of the thinking process, which can guide the pictorial elements of thinking into any logically permissible structure.

§ 14. GEOMETRY AS A THEORY OF RELATIONS

We must therefore reject the arguments for the priority of Euclidean geometry within mathematics. The geometrical axioms are not asserted to be true within mathematics, and mathematical geometry deals exclusively with implications; it is a pure deductive system.

We have based this assertion on our demonstration of the existence of visual pictures for both kinds of geometry and we must now investigate the extent to which visual pictures are actually necessary in mathematical geometry. The role they play in physical geometry is clear. They establish the relation between thinking and reality; they connect perceptions with concepts and therefore are involved in the important decision within physics which of the conceivable geometries corresponds to reality. There is no analogue for this decision in mathematics, because here no problem of choice among geometries exists.[1] What purpose do visual pictures serve?

To answer this question we must follow the theory of implicit definitions which was developed in connection with Hilbert's axioms.[2] According to this theory there is no need for visual pictures in mathematical geometry, and the mathematical meaning of geometrical structures and laws is exhausted by purely conceptual relations. The geometrical elements point, plane, line, etc., have no meaning other than that which is determined by their properties as formulated in

[1] We must also reject the epistemological equivalence between perception in physics and visualization in mathematics.

[2] See also the presentation by Schlick, *Allgemeine Erkenntnislehre*, Berlin 1918, pp. 30f. Our criticism of the theory of visualization corresponds in many respects to the presentation by Schlick (*ibid.*, p. 297), whose investigation of this question has become fundamental. For a comprehensive presentation of space as a relational structure which is particularly valuable because of well-chosen examples, see also Carnap, *Der Raum*, Chapter I (Ergänzungsheft der Kantstudien 1922).

the axioms. In mathematics, there is no additional visual significance.

We must say, for example: a point is something which can never lie on two different nonintersecting straight lines at the same time and which furthermore has the order property that between any two such things there is always at least a third, etc. In short, it satisfies all the conditions expressed in the axiomatic system. It is impossible, of course, to express this definition explicitly so that the defined concept of point no longer occurs on the right-hand side. We must speak therefore of an *implicit definition.* Let us consider an example from the realm of equations. The equation

$$y^2 = x + 2y$$

can be solved for y, in which case

$$y = 1 \pm \sqrt{1+x}$$

where y does not occur on the right-hand side. In contrast, the equation

$$x = \sin y$$

is implicit for y, i.e., it cannot be solved for y. Sometimes it is written in the apparently solved form

$$y = \text{arc} \sin x,$$

but the meaning of the function arc sin is defined only by the previously given implicit equation. It is therefore merely a restatement and not a solution by means of independently defined functions. This kind of empty restatement in an explicit form is possible also for implicit definitions. It appears already in the formulation: a point is something which has the properties determined by the axioms of Euclidean geometry.[1]

But does not this process define only the "objects" of geometry—do not the connecting relations *between, lies on,* etc., retain their visual significance? If this objection were justified, little would have been gained by the implicit definition of the basic elements. However, the relations too can be defined in this fashion. We can say: the relation *between* is the three-place relation b which applies to three points on a straight line and whose properties are to be determined. Thus the content of the different relations too is implicitly defined. This is not a circular procedure. What remains as undefinable basic concepts are such purely logical concepts as *element, relation, one-to-one*

[1] Compare this with the restrictions given on p. 97.

correspondence, *implication*, *and*, etc. All geometrical concepts, the elements as well as the relations, can be given as functions of these basic concepts.

This situation would appear more clearly if the axioms were not expressed in words but in logical symbols. Hilbert [1] has chosen words for the sake of greater lucidity because we find it difficult to understand the meaning of the basic concepts of geometry without visual pictures. His formulations, however, have been carried through so rigorously that a translation into logical symbols would be easy. This rigor constitutes the great significance of Hilbert's axiomatic system, which distinguishes it from the work of all of his predecessors. Especially the axiomatic system of Euclid is in this respect quite insufficient: he did not succeed in a complete conceptualization of the representing visual elements. To illustrate the purely logical meaning of Hilbert's axioms and to give an example of the symbolism of mathematical logic at the same time, we cite the three "between axioms" of Hilbert in this kind of language. The axioms that define the concept between are stated as follows: [2]

II, 1. If A, B, and C are points on a straight line, and B lies between A and C, then B lies also between C and A.

II, 2. If A and C are two points on a straight line, there exists at least one point B which lies between A and C and at least one point D such that C lies between A and D.

II, 3. Among any three points which lie on a straight line there exists exactly one which lies between the other two.

For our translation we need the following symbols: [3]

	Logical symbols		*Geometrical symbols*
	and (conjunction)	p(x)	x is a point
$\vee$	or (disjunction)	s(x)	x is a straight line
$\supset$	implies (has the consequence)	l(x, y)	x lies on y (the relation *lies on*)
(x)	for all x . . .	b(x, y, z)	x lies between y and z (the relation *between*)
$\exists x$	there is at least one x such that . . .		
$\bar{p}$	it is not the case that p		

[1] D. Hilbert, *op. cit.*

[2] *Op. cit.*, p. 6.

[3] For further explanation of logical symbolism, see the author's *Elements of Symbolic Logic*, New York 1947, §§ 6–7, 17–18.

The three axioms now read:

$$\text{II, 1.}\quad (x)(y)(z)\{p(x).p(y).p(z).(\exists w)[s(w).l(x, w).l(y, w).l(z, w)] \\ .b(y, x, z) \supset b(y, z, x)\}$$

$$\text{II, 2.}\quad (x)(z)\{p(x).p(z).(x \neq z).(\exists w)[s(w).l(x, w).l(z, w)] \\ \supset (\exists y)[p(y).b(y, x, z)].(\exists v)[p(v).b(z, x, v)]\}$$

$$\text{II, 3.}\quad (x)(y)(z)\{p(x).p(y).p(z).(\exists w)[s(w).l(x, w).l(y, w).l(z, w)] \\ \supset [b(y, x, z) \vee b(x, y, z) \vee b(z, x, y)] \\ .\overline{[b(y, x, z).b(x, y, z)]} \\ .\overline{[b(y, x, z).b(z, x, y)]} \\ .\overline{[b(x, y, z).b(z, x, y)]}\}$$

These sentences should be understood as follows: Only the logical symbols have an independent meaning; the geometrical symbols have a derived meaning—they denote elements and relations such as to satisfy the axioms, whose meanings are determined by the logical symbols alone.

We can easily show that the axioms entail a restriction such that only a certain number of the visual relations of geometry satisfy the axioms. Fig. 14 gives us the picture of three points on a straight line.

Fig. 14. The relation *between.*

Which of them lies between the other two? If we suppose that it is *z*, we have identified *b* with the visual relation ordinarily called "lies outside." Axioms II, 1 and II, 2 are compatible with this supposition, but not II, 3, since our assumption would give us:

$$(x)(y)(z)\{p(x).p(y).p(z).(\exists w)[s(w).l(x, w).l(y, w).l(z, w)] \\ \supset [b(x, y, z).b(z,x, y)]\}$$

This expression contradicts the symbolic expression of II, 3 given above, according to which only one of the expressions on the right of the horseshoe is permissible. The system defined by the three axioms therefore excludes our supposition. The same, of course, applies to the supposition that *x* lies in the middle. Our only choice is to consider *y* as lying in the middle. Only the visual relation *between* and not the visual relation *lies outside* satisfies the relation *b* which was defined exclusively by logical symbols.

Have we herewith determined the visual relations uniquely, i.e., defined them completely? By no means, since we know that other elements and relations satisfy the system of the Euclidean axioms as well. There are, of course, certain restrictions on the identification of the basic geometrical concepts with visual relations, but there always remains an arbitrary factor. The visual elements cannot be exhaustively defined by the basic logical concepts alone.

Let us recall the number system. If we understand by a "point" a triplet of numbers, i.e., an ordered combination of three real numbers, and by a "straight line" a linear equation, etc. . . ., the system of these "things" also satisfies the Euclidean axioms. This fact is the basis for the possibility of analytical geometry. We have been so conditioned to this way of thinking that the visual difference between the terms "straight line" and "linear equation" is often ignored in the same text and they are used interchangeably. This situation results from the fact that the properties of these concepts determined by the Euclidean axioms are the most important and the most frequently used. In other connections, however, the difference becomes obvious. A taut string can be identified with the visual picture of a straight line, not with that of a linear equation. The latter is coordinated to it, but is not visually equivalent.

Another example of elements that satisfy the Euclidean axioms is a set of elements which themselves belong to Euclidean geometry but which have an entirely different meaning from that of the words "point", "straight line", etc. Consider the well-known duality in projective geometry whereby the elements point and line can be interchanged. These elements have quite different visual structures but identically the same properties in projective geometry. A system that satisfies all the Euclidean axioms is formed by the so-called family of spheres. Imagine a fixed point P in space through which pass spheres and circles, such that P lies on the surface of the spheres and the circumference of the circles. Let us coordinate the following concepts:

a point: any point in space except P
a straight line: a circle through P
a plane: a spherical surface through P

while the other geometrical concepts retain their original visual significance relative to the new elements. For instance in the relation *between two points* the "between" now means along a circle and not along a straight line. This new system of elements satisfies all the Euclidean axioms without exception. This can easily be proved.[1] The axiomatic system, therefore, cannot exhaust the visual content of its elements and its relations. It cannot even uniquely determine a set of geometrical elements among the Euclidean entities.

Does this fact prove that mathematics cannot replace its visual components by logic? This interpretation would be erroneous. Mathematics does not deal with visual points and straight lines but only with the logical structures as defined by the axiomatic system. It is not the task of mathematics to coordinate visual pictures to these concepts; this task would lead from mathematics to physics.

If we say that mathematics can define its elements by means of implicit definitions, some qualification is required. R. Carnap directed my attention to this fact. Whether something is a point is not determined by its nature alone but depends on the other things to which it is to be related. One must always begin with a multitude of things and state which are to be points, straight lines, etc. Only then can one decide whether these things in their totality correspond to the coordinated concepts. Carnap calls such concepts as *point, straight line,* etc., which are given by implicit definitions, *improper concepts.* Their peculiarity rests on the fact that they do not characterize a thing by its properties but by its relation to other things. Consider for example the concept of *the last car of a train.* Whether or not a particular car falls under this description does not depend on its properties but on its position relative to other cars. We could therefore speak of *relative concepts,* but would have to extend the meaning of this term to apply not only to relations but also to the elements of relations. See Carnap, *Symposium* I, 1927, p. 355.

Why should mathematics use visual pictures? For the content of its assertions and for its logical conclusions it needs only the logical properties of these elements. The precision of mathematical reasoning lies specifically in the fact that it utilizes only the logically formulated properties of the visual structures. Visual structures are nothing but an aid to thinking and belong to the psychological apparatus which draws the conclusions, not to the content of the thoughts themselves. Thinking does not aim at the pictures but at the logical structure which they express. The psychological significance of an example rests on the fact that logical operations are facilitated when we think

[1] By means of stereographic projection (Fig. 9, p. 69) we can coordinate each spherical surface to a plane of projection. Each circle through P on the spherical surface is coordinated to a straight line in the corresponding plane of projection. This proves our assertion if we consider arbitrary spheres through P. It is evident that in the above table a complicated function must be substituted for the congruence of the Euclidean metric. See Weber-Wellstein, *Enzyklopädie der Elementarmathematik,* Teubner 1905, Vol. II, pp. 34f and 52f.

of concrete objects. If we write one of the logical arguments in its scholastic form: [1]

$$\begin{array}{c} M\ E\ P \\ \underline{S\ A\ M} \\ .\ .\ . \end{array}$$

it is very difficult to supply the correct conclusion, which in this case is E. (This is the so-called form *Celarent.*) However, if we now formulate a concrete example

No mammals have gills
All dogs are mammals
. . . .

the conclusion "No dog has gills" follows without effort. In this example therefore we perform nothing but a logical operation, whose logical structure is given in symbolic form, but whose manipulation is considerably facilitated by the logically inessential terms "mammals", "gills", etc. Similarly, it is easier for the mathematician to reach conclusions from the axioms if he imagines them realized by physical objects. His visual geometrical figures actually lead him into physics, not for physical purposes, however, but for the sake of the logical structure that is illustrated by the physical objects. This procedure does not make him a physicist any more than our performance of the inference *Celarent* makes us zoologists.

As a crude example let us consider the customary presentation of the vector calculus, which is usually developed as the physics of hydrodynamics. The purely mathematical concept of the *divergence* is introduced as source, and the *gradient,* in some cases, as velocity. While the mathematician is well aware in this case that he is using physics as a means of visualization, he forgets it in the usual visual representation of the geometrical axioms. Here, too, he is concerned with physics, namely the physics of rigid bodies and light rays. *"Pure visualization" means supplying the structure of mathematical relations with physical content analogous to supplying content to the vector calculus in terms of hydrodynamics.* In several branches of geometry this

[1] The symbols mean: S = subject term, M = middle term, P = predicate, A = universal affirmation, E = universal denial, I = particular affirmation, O = particular denial. If the two premises are arbitrarily given in their logical form, the conclusion is determined. The symbols A and E of the premises have already determined the corresponding symbol of the conclusion. Logical considerations determine what that symbol should be. In this case it is E. See the presentation of logic by I. v. Kries, *Logik*, p. 662, Tübingen, 1916.

procedure becomes evident. Let us recall the "construction of geometry under the presupposition of motions" as it was presented by Klein.[1] He speaks there about the *translations* of the space, considers all points shifted so that they coincide with certain other points, and formulates the properties of this shift by means of axioms. One of these axioms states that two translations can be interchanged (Fig. 15). We can move point A first sideways to A' and then up, or we can move it first up to A'' and then sideways. In either case we arrive at the

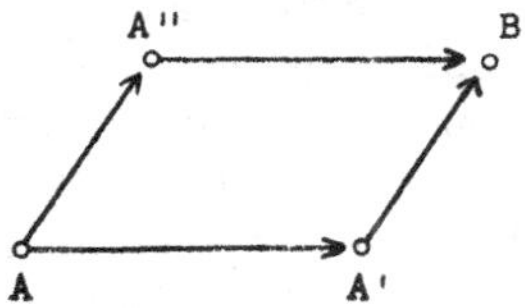

Fig. 15. The interchangeability of two translations.

same point B. It is obvious that dealing with the concept of motion does not represent a mathematical operation; a mathematical point cannot be moved, it can only be coordinated to another point. In the logical language of mathematics we should speak of a *coordinative operation* O which coordinates to the point A first a point A' and then coordinates to A' a point B. The coordinative operation between A and B is the *relative product* O_1O_2, and the axiom which we are considering states that the relative product O_1O_2 equals the relative product O_2O_1.[2] This product therefore satisfies the commutative law. In symbols, we have

$$O_1O_2 = O_2O_1 \quad \text{or} \quad AO_1O_2 \ldots = AO_2O_1 \ldots$$

This is the logical significance of our assertion. If it is formulated in the language of motion, and one speaks of shifting point A to point B, the logical framework is translated into visual pictures derived from the behavior of physical bodies moved along rigid rails. Such a translation does not change the mathematical relationships, nor does it give a different content to the mathematical assertion, but makes it more vivid and easier to understand. Although we must grant to the

[1] F. Klein, *Elementarmathematik vom höheren Standpunkte aus*, Vol. II, p. 174, Berlin 1925.

[2] The relative product is the relation resulting from the "arrangement in series" of two relations. The relation *brother-in-law* is the relative product *brother-spouse* (or also *husband-sister*).

mathematician the right to use such visualizations for the purpose of facilitating his thought processes, we cannot admit that it has any mathematical significance whatsoever.

The same arguments hold for the visualization of congruence by means of the superposition of line-segments. The logical significance of this motion of superposition is again a coordination of stationary line-segments. The visualization by means of transportation is derived from the experience with rigid bodies. We merely give some physical content to the logical framework. The concept of motion has no more significance here than had the zoological concept *animal* in the foregoing example of a logical inference.

We must therefore maintain that mathematical geometry is not a science of space insofar as we understand by space a visual structure that can be filled with objects—it is a pure theory of manifolds. In it, visualization plays the same role it does in arithmetic or in analysis; and, like the latter, it is reducible to basic logical concepts, namely the concepts of coordination, classes, etc., which constitute the actual content of geometrical assertions. The geometrical axioms are completely formulated as mathematical laws by formulae like those given on page 95. The visual elements of space are an unnecessary addition. Therefore the question of the truth of an axiom does not arise in mathematical geometry. Axioms are arbitrarily fixed relations, the content of which can be expressed by certain combinations of logical concepts alone, and which can be replaced just as well by any other consistent combination of basic concepts.

If we wish to express our ideas in terms of the concepts *synthetic* and *analytic*, we would have to point out that these concepts are applicable only to sentences that can be either true or false, and not to definitions. The mathematical axioms are therefore neither synthetic nor analytic, but definitions. This statement might be construed as a contradiction of our assertion that the visual compulsion of the geometrical axioms is logical in nature, because this assertion seems to indicate that the axioms are analytic. The seeming contradiction is resolved as follows: if a is a geometrical axiom, then points, straight lines, and other concepts which it may contain have no independent meaning. They obtain their meaning from a only in connection with the other axioms. We could consider the concepts point, straight line, etc., to be defined by the axiomatic system and to be reintroduced into axiom a. This procedure, however, would give us a new axiom which we might designate by a'; a' is analytic and true, whereas a is a definition and neither true nor false. Pure visualization gives us axiom a'; the geometrical axioms themselves are of the type a.

Hence the question whether axioms are *a priori* becomes pointless since they are arbitrary.

Strangely enough, F. Klein, who has made outstanding contributions to the development of non-Euclidean geometry, does not regard the axioms as arbitrary.

He calls them "not arbitrary, but reasonable statements which generally are evoked by spatial visualization and whose individual content is regulated on the basis of convenience" (*op. cit.*, p. 202). Nevertheless, the axioms of non-Euclidean geometry likewise are supposed to belong to these reasonable statements, since visualization can require the axiom of the parallels only within certain limits of accuracy (*op. cit.*, pp. 191 and 201). This point of view is untenable, since non-Euclidean geometry would become "unreasonable" beyond a certain degree of curvature. Klein's comments should be understood as a hint to mathematicians to think visually, rather than as an epistemologically conceived argument.

Only one system of axioms within mathematics retains its claims to truth, namely the axioms of logic itself. This system seems to be irreducible. An investigation of this question would lead far beyond the scope of this book, which deals with the problems of space and time. We can thus treat only the geometrical aspects of mathematics and shall be satisfied in having shown that there is no problem of the truth of geometrical axioms and that no special geometrical visualization exists in mathematics.

We have restricted ourselves in our formulation of the problem to geometrical visualization in order to avoid the discussions that have recently arisen in connection with mathematical intuitionism. Even though it is not possible to eliminate the visual element in mathematics altogether, there does not exist a special kind of geometrical visualization. It may be that intuitive visual processes enter into every instance of logical thinking and appear thus in all branches of mathematics in equal fashion. This question deals with the epistemology of logic and the relation between logic and mathematics, not with the specific problem of geometry and geometrical visualization. Above all, the results of such an investigation can never establish a difference between Euclidean and non-Euclidean geometry. There can be no doubt that the two systems occupy parallel positions in mathematics. When Hilbert (*Math. Ann.* 95, pp. 170–171) in his emphasis on the visual character of practical logical reasoning, believes that he confirms Kant's theory of visualization, he appears overly tolerant in his interpretation of an historical philosophical system. Kant's theory of the visualization of mathematics was based on the visual character of the synthetic axioms and was not concerned with the visual compulsion of analytic judgments. Hilbert's further remarks regarding the existence of extra-logical objects and the possibility of their visualization are relevant only to visualization in general. See also p. 107.

§ 15. WHAT IS A GRAPHICAL REPRESENTATION

The analysis of geometry as a theory of relations will become quite explicit if we use it to clarify a problem seldom recognized in its full scope. This is the problem of graphical representation.

Graphical representations are widely used. Every physical or technical text is filled with drawings of curves which enable us to understand the most complicated phenomena. The engineer cannot design his steam engines and motors, his bridges and electric circuits, without the use of diagrams; he needs these diagrams to calculate the efficiency of machines, to represent the strength of materials, and to obtain the stability of the vibratory state of motors and radio transmitters from the intersection of two straight lines on the diagram. These figures, whose dimensions he measures on graph paper and whose area he calculates with a planimeter, are aids not only to his understanding but also to his calculations. The physicist represents laws of nature graphically on the one hand by translating certain mathematical functions into curves and planes according to the rules of analytical geometry, on the other hand by plotting his measurements on graph paper and by connecting the points through curves in order to discover the represented functions. What the physicist represents as coordinates on graph paper is by no means only distances and spatial magnitudes. They can be pressures, temperatures, electric voltages, in short any measurable magnitude found in physics. This is all so well known that it need not be emphasized any further. Everybody has at one time or another welcomed the clarifying effect of a graphical representation while attempting to understand a physical problem.

We must ask the question, however, how it is possible that things so different as the gas law, the path of an electric discharge, etc., can all be represented by spatial diagrams. What do they have in common? Why, indeed, do such diagrams lead to an easier understanding? This fact is really quite strange. Would not the operator of a steam engine do a better job if he were thinking visually about the current of steam and the growing of the pressure as such? He could visualize them directly since he is able to perceive current and pressure. But does he use this kind of visualization? No, he looks at the manometer which provides a graphical representation of the pressure, and he estimates the amount of coal to be added by the amount which a pointer on a scale is short of the prescribed level.

If we were able to look into the minds of all these mechanics, electricians, and engineers, as we are able to look at a moving picture, we would find no images of pressures, voltages and lights, but invariably the sketch of a black curve on graph paper. Just think of a physicist in front of an electric switch; why does he turn the handle or move the wire into a certain position? Because his inner eye sees curves that increase, intersect, or decrease, and points traveling along a curve directed by means of switches. By far the most frequent visualizations of physical happenings are representations in terms of spatial relations that completely replace direct pictures. How is this possible?

The solution to this problem is contained in our conception of geometry as a theory of relations. The control of natural phenomena is achieved by means of mathematical concepts. These concepts are defined by implicit definitions and are not dependent on a unique and specific kind of visualization. Whatever visual objects we wish to coordinate to them is left to our choice. They may be pressures and currents as well as rigid measuring rods. This process of coordination is equivalent to a *coordinative definition.* There exists a coordinative definition not only for straight lines and rigid measuring rods, but also for straight lines and direct currents, or increases in the tension of a stretched rod. The coordination is arbitrary not only relative to certain kinds of things but also to the total domain of objects. The geometrical axioms can therefore be realized by means of compressed gases, electrical phenomena or mechanical forces as well as through rigid bodies and light rays. All these areas have a logical structure of such a kind that they can be coordinated to *mathematical* geometry; therefore they can also be coordinated to *physical* geometry and represented by means of diagrams.

Is a graphical representation actually a coordination to physical geometry? Do we coordinate here anything but *ideal* structures? We do indeed coordinate physical things, but it is somewhat difficult to notice this fact. We are so accustomed to the coordination of rigid bodies to mathematical geometry as a theory of relations that we no longer notice that there exists a duality. Nevertheless it is a coordination. On the one hand we have the mathematical system A of relations and on the other hand the physical system a of rigid bodies. Every assertion about A can be translated into an assertion about a, and it is customary to use assertions about a alone which are symbolic of assertions about A. This is called *visual* geometry. The system a is the *visual space* of A. In contrast, the content of A cannot be visualized and may be expressed by formulae like those given on page 95. This consideration also clarifies the term *pure visualization.* We do not think of the system a as a system of natural objects, but of objects exemplifying the relations of Euclidean geometry; then the system a of things is a *space of pure visualization.* Of course, we are not tied to a Euclidean geometry A but could choose a non-Euclidean geometry A' just as well. If we think of ideal objects a' which look like rigid bodies but satisfy the laws of A', then a' is the space of pure visualization of non-Euclidean geometry. The system a is a space either of pure or of empirical visualization, depending on whether we just invent the objects in a or find them in nature.

The so-called visual geometry is thus already a graphical representation, a mapping of the relational structure A upon the system a of real objects. It is therefore possible to represent even purely logical structures in graphical form. The representation of the inference of the scholastic schema (page 98) is the graphical representation of a logical inference. The logical relation of major premise, minor premise and conclusion is represented graphically by the spatial arrangement of the three lines. Another example of this type is the graphical representation of complex numbers, which were originally defined in purely logical terms, by means of points on a plane. This example shows how the systems A and a are merged in customary linguistic usage. Mathematics concerns itself with the coordination of the complex numbers to a system A, i.e., to a plane defined by the geometrical elements in the formulae on p. 95. Instead, one usually speaks of a coordination to the plane of a drawing, i.e., to the corresponding elements of a. Both terms are used interchangeably; thus we have the advantage of being able to think in pictures, since only the system a can be visualized, not the system A.

We can now understand why graphical representations are possible. The system of relations A can be coordinated not only to the physical system a, but to a large number of different physical systems $b, c \ldots$, for instance to the system of thermodynamics, to that of electrical phenomena, etc. Let us take as an illustration the P-T diagram of a gas of fixed quantity. The coordination for some of its elements is given in the following table: [1]

1. point	= the state of a gas at constant pressure and temperature.
2. straight line through the origin	= change of state at constant volume.
3. straight line parallel to the T axis	= change of state at constant pressure.
4. straight line parallel to the P axis	= change of state at constant temperature.

[1] The following expressions are justified by the gas law $PV = RT$.

5. any straight line	= change of state where the volume V is related to the temperature by the function[1] $V = \frac{T}{\alpha T + \beta}$
6. two parallel straight lines	= two changes of state as given in 5 with the same α and different β.
7. two equally long line segments	= two changes of state as given in 5 for which the expression $\sqrt{(P_2 - P_1)^2 + (T_2 - T_1)^2}$ depending on the initial and final states, has the same value.

On the basis of this coordination, the system *b* of the states of a gas is as good a realization of the system of relations *A* as the system of rigid bodies.

What do we do, however, if we are not satisfied with our table, but actually draw a diagram? We then carry out another coordination, namely that between *a* and *b*. We no longer speak of *A*, but of *a*. We say that in a change of state according to 5, the gas moves along a straight line, and we now understand by "straight line" a drawn figure and no longer a change of state. The straight line in the drawing is itself an object of *a* which we coordinate to the change of state. The connection between *A* and *a* is so close, that in order to understand the coordination between *A* and *b*, we establish the coordination between *a* and *b*. Of course, this is not logically necessary. We can omit *a* and treat *b* directly as a visualization of *A*. If we do this, however, we may no longer think of diagrams but must endow the objects in *b* with the qualities given in perception. We may think for instance of the sensation of pressure and temperature, any combination of both of which is a representation of a point. If we think of both as uniformly increasing, then the sensation which we have is the image of a straight line. The system *b* of states of the gas is as much a realization of *A* as the system *a* of rigid bodies and it is also a realization in terms of perceptions if we actually think of the states of a gas and do not smuggle in diagrams. There is not just one, but there are many different physical geometries. The geometry of rigid bodies is generally preferred for practical reasons, but it is by no means the only

[1] α and β are arbitrary constants.

one that can be visualized. We are so accustomed to the preferred position of rigid bodies that we acknowledge as space only the system represented by them. In principle, one could just as well call the manifold *b* of states of a gas a physical space. The "physical space" is generally reserved, however, for the system *a* of rigid bodies.

These practical considerations are naturally by far the most important ones and we shall therefore understand by physical geometry the system *a* of rigid bodies and light rays. This preference is caused not only by habit but is also based on the physical properties of the objects in *a*. They can easily be produced as geometrical tools such as rulers and triangles, easily be kept constant, and above all, readily be compared with other physical phenomena. These are the reasons why the objects in *a* have become the preferred tools of measurement and why we have grown accustomed to determine all physical states by comparison with rigid measuring rods. We measure temperature via a column of mercury finally on a rigid scale, an electric voltage by the length of a circular strip of paper, the scale of the voltmeter, etc. This preference, however, is not due to logical necessity. We could just as well measure the length of rigid rods by means of voltages if we were to substitute a standard dry cell for the standard meter kept in Paris. The reduction of all measurements to rigid measuring rods is based purely on pragmatic considerations. At various times attempts have been made to show the epistemological necessity of reducing all measurements to those of space (and time). This restriction is not admissible. The reduction is merely generally expedient because of the practical advantages of rigid measuring rods.

Now we understand the meaning of graphical representations. They signify nothing but a coordination of the system *a* to the systems *b*, *c*, . . . of other physical objects, which is possible only because all these systems are realizations of the same conceptual system *A*. That we call a graphical representation a visualization is epistemologically speaking not correct, since the systems *b*, c, . . . can be visualized just as well as *a*. Due to practical considerations, however, we have grown accustomed to express the conceptual relations of *A* mostly by *a*, not thinking of them abstractly but always as represented by visual pictures of *a*. It is a simplification if we represent other physical systems in terms of the preferred system *a*. Indeed, even the operator of a machine thinks more readily in terms of diagrams than in terms of the physical processes which he controls by the levers in his hand.

We should like to express at this point the conjecture that the representation of geometrical relations by systems of objects is more than a matter of convenience and that it rests on a basic necessity of human thinking. It is quite impossible to think abstractly about relations. We cannot understand them without some method of symbolic representation which supplies a concrete model of the abstract relations. The choice of system *a* is of course only one out of many possible selections. Even if we use the purely logical relations given by the formulae on p. 94, we are employing a concrete model when we think of the written letters, which are again nothing but a graphical representation of the system of relations. Thinking completely without symbols seems to be impossible. However, this fact should not lead to the mistaken impression that the chosen symbol is essential for the content of the thought. It is as irrelevant as is the color of the beads of an abacus for the arithmetical operations they represent. By content in the logical sense is meant only the system of relations common to a given set of symbolic systems. The fact that we can think of a system of relations only in terms of concrete objects does not change its independent and purely logical significance.

CHAPTER II. TIME

§ 16. THE DIFFERENCE BETWEEN SPACE AND TIME

Philosophy of science has examined the problems of time much less than the problems of space. Time has generally been considered as an ordering schema similar to, but simpler than, that of space, simpler because it has only one dimension. Some philosophers have believed that a philosophical clarification of space also provided a solution of the problem of time. Kant presented space and time as analogous forms of visualization and treated them in a common chapter in his major epistemological work. Time therefore seems to be much less problematic since it has none of the difficulties resulting from multi-dimensionality. Time does not have the problem of *mirror—image congruence*, i.e., the problem of the existence of equal and similarly shaped figures that cannot be superimposed, a problem which has played some role in Kant's philosophy. Furthermore, time has no problem analogous to non-Euclidean geometry. In a one-dimensional schema it is impossible to distinguish between straightness and curvature. A curved line can always be "straightened out" without a deformation of its smallest elements. It is therefore impossible to determine by internal measurements whether a one-dimensional continuum is straight or curved. A line can have an external curvature but never an internal one, since this possibility exists only for a two-dimensional or higher continuum. Thus time lacks, because of its one-dimensionality, all those problems which have led to the philosophical analysis of the problems of space.

The treatment of the problem of time as parallel to that of space has been detrimental. One was aware only of those problems which do t exist for time, rather than of its special features. These features manifest themselves in the fact that time order is possible in a realm which has no spatial order, namely the world of the psychic experiences of an individual human being. This is the reason why the experience of time is allotted a primary position among conscious experiences, and is felt as more immediate than the experience of space. There is indeed no experience of space in the direct sense in which we feel the flow of time during our life. The experience of time appears to be closely connected with the experience of the ego. "I am" is always equivalent to "I am now," but I am in an "eternal now" and feel myself remaining the same in the elusive current of time.

At the moment, however, we cannot go into this question. Before we attempt a solution of these intricate problems, it is necessary to consider the order of time as a problem of natural science, similar to that of the order of space. An analysis of natural science is the only path to the central problems of epistemology. We must therefore first examine those problems which result from the parallelism of spatial order and time order and show that the changes in the philosophical analysis of geometry have also consequences for the order of time. First, the problem of congruence exists for time intervals as well as for spatial distances. The parallelism becomes even closer, if space and time are combined into a four-dimensional manifold where all the epistemological problems appear in the same fashion in which we have encountered them in the three-dimensional manifold of space.

Whereas the conception of space and time as a four-dimensional manifold has been very fruitful for mathematical physics, its effect in the field of epistemology has been only to confuse the issue. Calling time the fourth dimension gives it an air of mystery. One might think that time can now be conceived as a kind of space and try in vain to add visually a fourth dimension to the three dimensions of space. It is essential to guard against such a misunderstanding of mathematical concepts. If we add time to space as a fourth dimension, it does not lose in any way its peculiar character as time. Through the combination of space and time into a four-dimensional manifold we merely express the fact that it takes four numbers to determine a world event, namely three numbers for the spatial location and one for time. Such an ordering of elements, each of which is given by four conditions (coordinates) can always be conceived mathematically as a four-dimensional manifold. The same is of course possible in many other cases. Musical tones can be ordered according to volume and pitch, and are thus brought into a two-dimensional manifold. Similarly, colors can be determined by the three basic colors, red, green, and blue, if we state for any given color how much it contains of each of these three components. Such an ordering does not *change* either tones or colors; it is merely a mathematical expression of something that we have known and visualized for a long time. Our schematization of time as a fourth dimension therefore does not imply any changes in the conception of time.

The practical value of this form of mathematical expression lies in the fact that we can occasionally visualize the manifold with the aid of spatial concepts, i.e., that we can represent them graphically. We can thus symbolize the manifold of tones by means of a plane. If we express the volume of the tone on a horizontal axis and its pitch on a vertical axis, then every point of the plane (more correctly, the quadrant, since volume and pitch cannot be negative) corresponds to a tone of specific volume and pitch. Such a representation of tones on a plane is for many purposes very practical, but it is by no means necessary. Even if we understand by volume and pitch the experience that we have in hearing the tone, the two-dimensional manifold still exists; these experiences themselves form the manifold. Let us refer at this point to the considerations of § 15 which showed that a multi-dimensional manifold is a conceptual structure and that the space of visualization is only one of many possible forms that add content to the conceptual frame. We therefore need not call the representation of the tone manifold by a plane *the* visual representation of the two-dimensional tone manifold. The auditory realization of the tone experiences themselves would also give perceptual content to the conceptual manifold. The same holds for the four-dimensional space-time manifold. We *could* conceive it as represented by a four-dimensional space; in this case, however, imagination fails us, since visualized space has only three dimensions. In this situation we can avail ourselves of spatial representations of cross-sections of the four-dimensional manifold. We may represent a dimension of space on a horizontal axis, the dimension of time on a vertical axis, and obtain in the plane of the resulting space a representation of the manifold of events which occur on a line in space at various times. This method of visualizing the flow of time by means of a diagram can be very useful. The theory of relativity, however, is not required for such a visualization, since graphically represented railroad schedules, for example, achieve the same effect.

Yet this device does not change our conception of time. We can always fill the four-dimensional space-time manifold with the direct perceptual content which we have connected with space and time in the past. In the way we experience events as spatially and temporally determined, they already form a four-dimensional manifold. We may therefore retain the perceptual difference between space and time without fear of contradicting the mathematical representation. Just as the representation of tones on a plane cannot require us to give up the intuitive representation of the volume or pitch of a tone, so the combination of space and time into a four-dimensional manifold cannot offer any grounds to discard the intuitive representations which we connect with space and time and which differ considerably for them. On the contrary, it is just in this form that the intuitive representation of all four dimensions is possible without difficulty.

The properties of time which the theory of relativity has discovered have nothing to do with its treatment as fourth dimension. This procedure was already possible in classical physics, where it was frequently used. However, according to the theory of relativity the four-dimensional manifold is of a new type; it obeys laws different from those of classical theory. These results were obtained when time was subjected to the same kind of analysis as was applied to the three-dimensional space manifold. The analysis led to a realization of the arbitrariness of coordinative definitions even for time, and finally to some insights which appeared at first very strange. To demonstrate these changes in the conception of time we need not employ mathematical considerations. We can remain within the perceptual experience of time and develop everything the theory of relativity teaches about time. Indeed, in an epistemological sense we shall go further. On the other hand, we shall recognize the significance of the structure of the four-dimensional manifold with the help of a mathematical formulation of our results, and thus deprive it of its apparent mystery. We shall even find that the Minkowskian world is incorrectly interpreted if one looks to it for support of the parallelism of space and time; on the contrary, the world of Minkowski expresses the peculiarity of the time dimension mathematically by prefixing a minus sign to the time expression in the basic metrical formulae. The peculiarity of time appears even in an analysis that does not consider the subjective experience of time. We shall show that the parallelism does not exist objectively and that in natural science time is more fundamental than space, the topological and metrical relations of which can be completely reduced to observations of time. We shall finally recognize that time order represents the prototype of causal propagation and thus discover space-time order as the schema of causal connection.

In this chapter we shall consider only physical time. We shall pay no attention to the psychological characteristics of the experience of time, but shall analyze the physical order of time just as we gave an analysis of the physical order of space. Such a distinction is certainly possible; we can examine what physics means by "time" just as we can examine what physics means by "matter." It has often been claimed that only the physical properties of time can be revealed in such an investigation and that, unaffected by physical time, the psychological experience of time retains its *a priori* character and obeys its own laws. This view which has been expressed by various philosophical writers in connection with the theory of relativity, must be rejected most emphatically. All our so-called *a priori* judgments are determined by primitive experiences, by the physics of everyday life, to a much higher degree than we may think. Nothing would do more harm to the progress of science than to interpret such experiences as apodictic necessities and thus to arrest the natural growth of our knowledge. Actually, such a conception would make the physics of everyday life the norm for scientific physics and express our unwillingness to adjust our imagination to the development of physics from a naive world picture to an exact science. We shall therefore use the distinction between *time as experience* and *physical time* only as a temporary aid which leads us to a deeper scientific insight into the concept of time; we shall correct the intuitive experience *time* accordingly. Indeed, we shall find that it is just the relativistic concept of time which presents the experience of time in a new light. This analysis will clarify the meaning and content of everyday experiences; finally we shall learn in this way, better than through a phenomenological analysis, what we "actually mean" by the experience of time.

§ 17. THE UNIFORMITY OF TIME

The solution we have offered for the problem of physical geometry is based on the idea of the coordinative definition. The first coordinative definition referred to the unit of length and the second to congruence. Whether two distant line-segments are equal is not a matter of *knowledge* but of *definition*; and this definition consists ultimately in a reference to a physical object coordinated to the concept of a unit. We recognized the need for such a coordinative definition because otherwise the problem would remain undetermined. It is not a technical but a logical impossibility to compare distant line-segments without a prior coordinative definition of congruence. The definition of congruence by means of rigid bodies proved to be most useful, since this definition was shown to be independent of the path along which the rigid body is transported.

Similar considerations must be carried through for the problem of time. It is so obvious that we have to determine a unit of time, that we shall merely mention this first coordinative definition. But for time, too, there is a comparison of length. Before we enter into an epistemological investigation, let us first examine what time intervals physics considers to be equal in length. The rotation of the earth is the most important example; we say that the time intervals which the earth requires for one complete rotation are equal. For the subdivision of such time intervals we use a different method, namely the measurement of angles. We accept time intervals as equal if they correspond to equal angles of the earth's rotation. Through the combination of these two methods we obtain the measure of time, and the flow of time we have thus obtained is called *uniform*. The problem of the congruence of time intervals leads therefore to the problem of the *uniformity of time*.

The described time measurement employs two essentially different methods. If we consider the revolutions of the earth to have equal duration, we do this because they represent *periods of the same type*. The same principle is involved if we say that the periods of a pendulum are equally long. The counting of periods is the first and most natural type of time measurement. The second method consists in subdividing the diurnal period by means of the angle of the earth's rotation. In this case, equal times are measured with the aid of equal spatial magnitudes. This reduction of time measurements to space measurements is also present in inertial motion. According to the law of inertia, if a body moves freely, unaffected by accelerating or retarding forces, it will cover equal distances in equal time intervals. We can thus use its motion as a measure of uniformity and regard as equal the times of transit through equal distances. Finally, the motion of light permits an analogous method since light covers equal distances in equal times. There are therefore two basic kinds of time measurement: one consists in counting *periodic processes*, and the other in *measuring spatial distances* corresponding to certain non-periodic processes.

The opinion has been expressed that there are no actual time measurements and that all time measurements must be reduced to spatial measurements. This is not correct. The reduction applies only to the second type of time measurement; the first method has nothing to do with spatial measurements. If we count periodic events, as for instance the tick-tock of a watch, we are using a genuine time scale. We hear a sequence of sounds and call the intermediate time intervals equal. That we call them equal is based on the fact that each sound represents the end of a full period at which the swinging pendulum has reached its previous position. How it moves within one of these periods does not matter at all. It is well known that the motion of a pendulum is far from uniform and yet we accept the intervals of the complete periods as equal. The fact that a period is completed is recognized by the return of the system to its original condition; there is no need of a spatial measurement. This time measurement is thus based on the recurrence of the same state. The watch is a good illustration of this procedure. The internal works have in this case only the significance of a *counting device,* and the angular path of the hands is merely a measure of the number of cogs which the gears have advanced and hence also a measure of the number of completed periods of the balance wheel. The time measure of the watch is therefore provided by the balance wheel; the hands merely indicate the *number of units of measurement* and save us the trouble of counting. Actually, we can only measure an integral number of time intervals by this method. If the unit is chosen small enough, however, the resulting inaccuracy can be made very small.

In special cases the individual period is run through uniformly as, for instance, in the case of the rotating earth. We arrive at a subdivision according to the second method by measuring the angular path of the earth's rotation relative to the fixed stars. This subdivision of the time measure involves genuine spatial measurements, namely, of angular distances, and differs therefore from the merely apparent use of angular measurements in the case of the watch.

To summarize our ideas, we may say that the measurement of equal time intervals is obtained through mechanisms which we assume to run through their periods in equal times. Actually we never measure a "pure time," but always a *process*, which may be periodic as in the case of the clock, or nonperiodic as in the case of the freely moving mass point. Every lapse of time is connected with some process, for otherwise it could not be perceived at all. The measurement of time is therefore based upon an assumption about the behavior of certain physical mechanisms.

How can we test this assumption? There is only one answer: we cannot test it at all. There is basically no means to compare two successive periods of a clock, just as there is no means to compare two measuring rods when one lies behind the other. We cannot carry back the later time interval and place it next to the earlier one. It is possible to make empirical statements about clocks, but such statements would concern something else. Two clocks stand next to each other, and we observe that the beginning as well as the end of their periods coincide. Further observation may show that the ends of their periods always coincide. This experience teaches us that two clocks standing next to each other and having equal periods *once* will *always* have equal periods. But this is all. Whether both clocks require more time for later periods cannot be determined.

Why is this determination impossible? Do not the laws of physics, for instance those of the motion of a pendulum, compel us to believe in the equality of the periods? It is true that the laws as described in textbooks suggest this belief; but if we ask ourselves where these laws come from, we shall find that they are obtained through observations of clocks calibrated according to the principle of the equality of their periods. The proof is therefore circular. If we had used a different scale for our measurements, we would have obtained different laws which in turn would have compelled us to consider the latter scale as the correct one. Neither can the circularity be removed by time measurements of nonperiodic processes. The law of inertia does prescribe a measure of time, but this law could easily be restated for a different type of time measurement in which a freely moving body slows down and a body which falls towards the earth moves at a uniform rate—this restatement would never lead to internal contradictions.

A solution is obtained only when we apply our previous results about spatial congruence and introduce the concept of a *coordinative definition* into the measure of time. The equality of successive time intervals is not a matter of *knowledge* but a matter of *definition*. As for spatial congruence, a certain rule must be laid down before the comparison of magnitudes is defined. This determination can again be made only by reference to a physical phenomenon; a physical process, such as the rotation of the earth, is taken as a measure of uniformity by *definition*. All definitions are equally admissible. We could define the motion in the earth's gravitational field as uniform and would consequently obtain a retardation for a freely moving body. Physics, however, has decided on a particular definition with special properties. It uses three independent methods for the definition of the uniformity of time:

1. The definition by means of natural clocks.
2. The definition by means of the laws of mechanics. (It comprises not only the definition by means of inertial motion, but also those definitions which use the rotating earth or the pendulum.)
3. The definition using the motion of light (light clock).

We shall discuss 1 and 2 in the following section, 3 in § 27. However, we can assert: it is an empirical fact that these three definitions lead to the same measure of the flow of time. Since these definitions have this property, the clock proves to be the natural measure of time in the same sense in which the rigid measuring rod is the natural measure of space.

Processes of nature thus determine a flow of time. It is, however, not an epistemological necessity to use the clock as a definition of uniformity. In an epistemological sense any other definition is equally admissible, provided only that it leads to a univocal and noncontradictory description of nature. For practical reasons one chooses the definition by means of clocks because it simplifies the description of nature considerably. This simplicity has nothing to do with truth, since it is merely *descriptive simplicity*.

On the other hand, it is a statement of fact that a flow of time of this kind exists; that therefore all periodic processes, and furthermore inertial motion and the motion of light, lead to the same measure of time. This statement should not be considered to be *a priori* but the result of experience. It could be false, and later we shall learn about cases for which it is indeed not true. Today we know that it applies strictly only in gravitational-free space and in gravitational fields of particular simplicity (in stationary fields). Since strictly speaking there are no such fields, our characterization of the uniform flow of natural processes holds only approximately.

This approximation fits terrestrial and astronomical relations to such a high degree that the deviations lie far below the limits of exactness. There is good reason, therefore, for astronomers to try to make uniform time independent of the fluctuations of the earth's motion resulting from its own rotation, axial oscillations, its revolution around the sun, and lunar influences. These difficulties show that the coordinative definition of uniformity cannot be given so easily as it can be schematically conceived. There is no periodic motion which is completely free from external influence and which returns to exactly equal states. Even the earth's rotation has these properties only to a certain, though very high, degree of approximation. The precession of the earth's axis has the effect that the earth has a slightly different position after each rotation and therefore does not reach exactly the same state. For this reason uniform time is not considered to be equal to the directly observed time, but is derived indirectly from it by a series of corrections. This method is the same as that used in the measurement of length, where the unit of length is given not directly by the transported measuring rod but is calculated indirectly with the aid of correction factors for temperature, etc. It is obvious, of course, that this method does not enable us to discover a "true" time, but that astronomers simply determine with the aid of the laws of mechanics that particular flow of time which the laws of physics implicitly define. A redefinition of uniformity through a change in the laws of physics would give the astronomer a different time. His work is comparable to the investigation of the physicist to determine the c.g.s. unit of electric current, if the ampere is already defined by the electrolytic separation of a specific quantity of silver. This is an exceedingly difficult task, which is of great importance, but it does not teach us how large the unit of current *should* be.

We can schematize the definition of uniformity given by the laws of physics in the same way as we schematized the definition of the comparison of length. For this purpose we introduced in § 6 the distinction between universal and differential forces. Universal forces are those that affect all substances equally, whereas differential forces affect them differently. We shall use the same distinction in our definition of the clock, which we defined above as a closed periodic system. However, the concept of a closed system is not defined so long as universal forces are permitted. If we should regard the period of the earth's rotation as variable—for example (starting from an arbitrary point) call the second rotation twice as long, the third three times as long—then this definition would become noticeable in the equations of physics through the appearance of a force which was thus introduced by definition. This force would have the "effect" that the period of rotation would constantly increase. We would find that this force retards all clocks in equal fashion and that it retards the motion of all otherwise freely moving bodies; it has all the properties of a universal force. We now set this force equal to zero by definition, i.e., we define the closed system as free from differential forces, but neglect universal forces. This definition therefore determines the zero point from which forces are measured. Without such a zero point the magnitude of a force would be left undetermined, since a force is something which we regard as the cause of a change, and a change of temporal or spatial intervals can be determined only if a coordinative definition of congruence has previously been given. For this reason the definition of the congruence of time intervals is connected with the problem of a force field. The definition of congruence for time comparison is therefore also the basis for the measurement of a force, and conversely this definition of congruence can be given through the rules for the measurement of a force.

We must finally recall another difficulty which exists for any definition of a closed system. We can never construct a system completely isolated from external differential forces, because this is possible only to a certain degree of approximation. Consequently we can define only the concept *closed to a certain degree of approximation.* This degree of approximation, however, depends on the relation between the external forces and the internal forces of the system. In a given field of external (differential) forces, one system can be relatively well closed, another relatively badly closed. Furthermore the same system can sometimes be relatively well closed, and sometimes relatively badly closed, depending on the external (differential) field.

§ 18. CLOCKS USED IN PRACTICE

Let us consider in this connection the clocks in actual use. There is the pendulum clock, the spring-balance clock (pocket watch), the earth clock, and finally the atomic clock of the revolving electrons within the atom. Actually, pendulum clocks are not clocks because they are not closed systems. They move only because of the earth's gravitational force, which belongs to the class of universal forces and could be eliminated by definition. Simultaneously, always we find a differential force that affects the supporting arrangement of the pendulum and compensates for the gravitational force. This elastic force is an external differential force which is essential, however, because if it did not exist, the pendulum would not oscillate but would fall freely. This force is of the same order of magnitude as the driving force of the pendulum and the system is thus far from being closed. The pendulum clock can therefore be used as a measure of uniform time only with certain precautionary measures, namely, when the force of attraction of the earth is constant. The pendulum clock is really nothing but an indicator of the earth's attraction, a force which is measured by the equations of mechanics, such that it is constant when time is defined by truly closed clocks, as for example the earth clock.[1] Therefore, even the pendulum clock may be taken as a measure of uniform time. If we were to move a pendulum clock to a different latitude, however, the same equations of mechanics would force us to consider the unit of time of the clock as changed, since the strength of the gravitational field of the earth is smaller near the equator than at the poles due to the considerable flattening of the earth. The definition of time by means of the pendulum belongs therefore to the second type of definition given in the previous section (page 117).

The spring-balance clock is a clock in our sense. The force that determines the length of the pendular period of the balance wheel is the elastic force of the spring, i.e., the *internal* force of the clock. Gravitation has no effect in this case, and the indications of the spring-balance clock are thus independent of latitude. This kind of clock would oscillate even if it were far away from any masses in interstellar space, whereas the pendulum clock would not operate under these conditions. We may therefore always think of a pocket watch, when we talk of clocks in our future discussions, since it is the best example of a closed system. For precise time measurements watches have shortcomings, however, which make them inferior to the pendulum clock for astronomical purposes. The elastic forces fluctuate slightly, i.e., the system is not strictly periodic, and therefore, although this system is closer to our epistemological ideal, its time is not as accurate as that of the pendulum clock. So long as pendulum clocks remain in the same place they satisfy the condition of uniformity to a very high degree, because the earth's attraction is very nearly constant. On ships, on the other hand, spring-balance chronometers are used, because the effect of geographical latitude would otherwise be noticeable, and the rocking of the ship would cause fluctuations in the

[1] It is easy to recognize that the measure of force depends on the measure of time which appears in the dimensions mlt^{-2} of force. A force is measured by the acceleration which it produces. If the measure of time were redefined in a suitable fashion, the acceleration of a freely falling body would not be constant and the gravitational force would vary with time.

elastic supporting forces which for pendulum clocks are of the same order of magnitude as the driving forces.

One should not forget, however, that the spring-balance clock too is only approximately free from external physical forces. It must, for instance, be placed on a firm support to prevent its fall in the gravitational field. This elastic supporting force, which is external, affects the clock and causes a slight bending of its gears and shafts. The spring-balance clock thus runs somewhat differently, depending upon the side on which it rests. This effect is extremely small, and we may say in our language: the external physical forces are very small in comparison with the internal driving forces of the clock. This type of clock is therefore closed to a relatively high degree of accuracy. This applies, of course, only so long as the elastic forces are not too large. If we were to fasten a pocket watch to a rotating disc, for instance, and bring it into a strong centrifugal field, the supporting forces would increase correspondingly and would bend the gears and shafts so much that the clock could no longer be considered a closed system.

The most important clock for practical time measurements is the earth clock. We have already mentioned that this clock too needs corrections because it is not strictly periodic, but it is far superior in this respect to the spring-balance clock. Since it floats freely in space and the effect of radiation can be ignored, the earth clock is a well-closed system. There is a slight disturbance by the gravitational effects of the moon and sun; they act as brakes upon the earth's rotation with the eventual result of a state like that of the moon, for which the period of rotation equals that of the orbit. For a single rotation, however, this effect can be ignored.

The earth clock has the characteristic that it moves uniformly even within the individual period. Thus the unit of time can be subdivided by means of angular measurements. On the other hand, the counting of periods is more difficult, since the end of an individual period cannot be directly recognized. What constitutes a complete rotation of the earth is definable only relative to the environment; hence the difference between stellar and solar day. The latter is about four minutes longer, because the period between two solar culminations increases due to the orbital motion of the earth. If the earth were alone in the universe, it would be useless as a clock, because we would have no indication for the end of its individual periods. The earth clock is therefore not a natural clock in our sense. In contrast to the pendulum clock it is closed, but it differs from ideal clocks in the sense that it is not periodic. The earth's rotation is an inertial motion, and the definition of a measure of time by means of the earth clock should therefore be included among the definitions that use the law of inertia rather than a periodic system.[1] Therefore, the question when a "real" rotation of the earth has been completed can be answered only on the basis of the laws of mechanics. The answer depends on the determination of the astronomical inertial system (compare § 36), to which the earth's rotation must be related for this purpose. Since the stellar day is practically identical with this "day of the inertial system," it is generally preferred to the solar day.

One might suppose that these considerations are unnecessary as long as we are interested only in the uniformity of time and not in the length of an individual day, since we can employ the method of subdivision by angular measurements because of the uniformity within a period of the earth's rotation. But how are we to measure this angle of rotation? It is well known that the angle of rotation measured relative to the sun is not suitable for this subdivision (because of the varying velocity of the earth along its path) but only the angle relative to the fixed stars. The only difference is the following: In order to determine uniformity it is sufficient to know the state of the astronomical inertial system with the exception of a uniform rotation, whereas for the determination of the length of a day even this variable must be eliminated. Practically speaking, however, little is accomplished by the admission of such an unknown rotation, since the determination of the inertial system except for a uniform rotation contains almost all the problems pertaining to the determination of the rotation-free inertial system.

We should finally mention the atomic clock, which, although of no practical use, plays an important role in the experimental investigations of the relativistic laws dealing with clocks. An electron within the atom revolves with a high degree of precision and its period provides us therefore with a very exact unit of time. The force that the nucleus of the atom exerts on the electron, keeping it in a closed path, can be considered as the internal force of the clock. The entire atom, and not the electron alone, is the clock the period of which is indicated by the revolution of the electron. This clock is closed to a very high

[1] The elliptic motion of the earth around the sun is a periodic process. The point at which the earth is closest to the sun, the perihelion, may be regarded as the end of a period. This, however, is not an inertial motion, but a gravitational motion.

degree since the external forces acting on an atom are very weak compared to its strong internal forces. The atom would be an ideal clock except for the results of quantum theory. We can never observe an atomic clock as we observe other clocks; we can only measure the frequency of the emitted radiation. According to the ideas of classical theory, the frequency would be a direct measure of the period of the electron's revolution, which could thus be directly observed. Bohr's discoveries have taught us, however, that the atom emits light in an entirely different fashion. The revolving electron does not emit any light at all; consequently we have no knowledge of its period of revolution. Light is emitted only when the electron jumps from one orbit to another; the conditions of a periodic system are, therefore, not satisfied. Bohr's theory has shown nevertheless that the frequency is determined by the stationary states between which the jump takes place. However this situation is much more complicated and the question arises whether under these conditions the atom may still be regarded as a clock. As long as we have no accurate information about the process of the emission of light, we can only express conjectures at this point. On the other hand one could, of course, investigate experimentally how far the atomic clock satisfies the relativistic laws of clocks. In this manner we might discover directly whether or not the atom can be regarded as a clock in the sense of the relativistic theory of time.

§ 19. SIMULTANEITY

After we had specified the unit of time, which is the first metrical coordinative definition of time, we were led to the problem of uniformity, which is the second metrical coordinative definition of time and deals with the congruence of successive time intervals. There is however a second type of time comparison that concerns parallel time intervals occurring at different points in space rather than consecutive time intervals occurring at the same point in space. The comparison of such time intervals leads to the problem of simultaneity and hence to the third metrical coordinative definition of time. Although it had been known for some time that uniformity is a matter of definition—Mach,[1] for instance, asserted the definitional character of the uniformity of time emphatically—the definitional character of simultaneity was

[1] E. Mach, *The Science of Mechanics*, The Open Court Publishing Co., Chicago and London, 1919, p. 223.

recognized first by Einstein and has since become famous as the relativity of time. Einstein immediately applied his solution of the problem of simultaneity to theoretical physics and for this reason the epistemological character of his discovery has never been clearly distinguished from the physical results. Therefore, we shall not follow the road taken by Einstein, which is closely connected with the principle of the constancy of the velocity of light, but begin with the epistemological problem.

To see this problem clearly, we must start with a distinction which originated with the work of Einstein. We shall distinguish between the *simultaneity at the same place* and the *simultaneity of spatially separated events.* Only the latter contains the actual problem of simultaneity; the first is strictly speaking not a simultaneity of time points, but an *identity.* Such a concurrence of events at the same place and at the same time is called a *coincidence.* In a strict coincidence there is actually no comparison of space or time since position and time are identical for both events. Practically speaking, such an identity never occurs since we could no longer distinguish the two events. But an approximate coincidence can be realized, in the example of two colliding spheres or two intersecting light rays. Simultaneity plays no essential role even in the case of a roughly approximated coincidence, because a time comparison of distant events shows such slight differences in the determination of the time of neighboring events that they can be ignored. We can therefore treat the problem of *the comparison of neighboring events* similarly to the problem of coincidence and restrict our investigation to the *comparison of distant events.*[1]

This investigation will lead us to the result that the simultaneity of distant events is based on a coordinative definition. We shall demonstrate this result by showing that a comparison of time has the characteristic properties of a coordinative definition. We therefore maintain:

First, it is impossible to ascertain whether two distant clocks are set "correctly" in their indication of time; second, they can be set arbitrarily and yet no contradiction will arise.

Following the first line of thought, we may ask how one can determine the simultaneity of distant events. We shall consider events as distant, if the distance between them is large compared with the dimensions of the human body. The perceptual judgment of simul-

[1] For a rigorous treatment of the comparison of neighboring events, see A., § 8.

taneity is thus not sufficient under these circumstances. We may hear for instance the sound of thunder and notice at the same time that the hands of our watch point to 8:50. The determination of simultaneity which we make here is a comparison of neighboring events; we compare the moment when the watch indicates 8:50 with the moment when the sound of thunder reaches our ear and not with the instant of its occurrence. If we want to derive from this time determination the actual time at which the thunder occurred, we must have additional physical facts. We must know the distance which the sound has traveled and the velocity of sound, before we can calculate backwards from 8:50 to the time at which the thunder took place.

But are there no other means? It is well known that we can avoid using the velocity of sound in the given example, if we observe the lightning rather than the thunder. Let us say that the lightning was observed at 8h 49m 50s; we may then consider this time as the time at which both lightening and thunder occurred. Is this statement true? Obviously, in this type of time determination the situation is changed quantitatively but not in principle. The light of the lightning also requires a certain amount of time to reach the eye, and our judgment therefore concerns again the moment at which the light reaches our eye and not the moment when the lightning actually occurred. Only because this time difference is extremely small can we ignore it for practical purposes.

It can easily be seen that the time comparison of distant events is possible only because a *signal* sent from one place to another is a *causal chain.* This process leads to a coincidence, i.e., a comparison of neighboring events, and from the time measurement thus obtained we can determine the time of the distant event only with the help of an *inference.* What assumptions are contained in this inference?

This inference requires besides the knowledge of the distance also the knowledge of the velocity of the signal. How can this velocity be measured?

In principle, there exists only *one* method, which we shall schematize as follows. The signal leaves a point P_1 at the time t_1 and reaches a point P_2 at the time t_2. Its velocity is then given by the quotient of the time interval t_2-t_1 and the distance P_2-P_1. Therefore, *two* time measurements are required which have to be made at *different* places. We can think of them as given by two clocks located at P_1 and P_2. If the indication of the time interval t_2-t_1 is to be meaningful, however, the two clocks must have been synchronized previously, i.e., it must have been determined whether their hands occupied the same positions at the same time. In order to measure a velocity, therefore, the simultaneity of distant events must already be known.

Is this statement correct? Did not Fizeau measure the velocity of light differently? Fizeau indeed used an arrangement which did not require the simultaneity of distant events. We can schematize his measuring arrangement as follows. In Fig. 16 a light ray is sent from

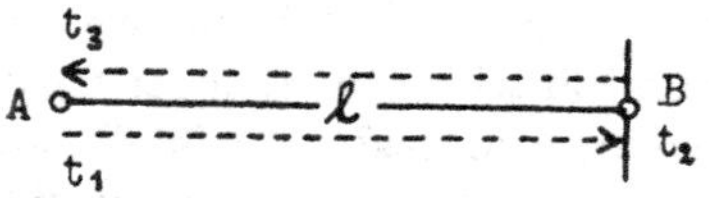

Fig. 16. Round trip of a light signal.

A at the time $t_1 = 12{:}00$; it is reflected at the point *B*, which is at a distance l from *A*, and finally returns to *A* at $t_3 = 12{:}06$. It has required 6 minutes to travel twice the distance l, and its velocity is thus given by the ratio of these two numbers. In this arrangement, time is measured only at *A*, only *one* clock is used and the simultaneity of distant events does not affect the problem. Of course, in the actual experiment the time interval was much smaller than 6 minutes, even though l was several kilometers long, but Fizeau measured it by an ingenious device involving a rotating gear. We have simply chosen larger numbers to clarify the illustration.

On closer examination, we notice that this measurement contains a certain untested assumption, namely, that the velocity of light is the same in both directions along l. For instance, if it were less, in the direction *AB* than in the direction *BA*, the velocity of light as calculated by Fizeau would correspond to neither of the two velocities, but represent an average of the two. How can we prove this assumption by Fizeau?

It seems that it can be proved only if the time t_2 is known at which the light ray reaches *B*. This means, however, that we are again employing *two* clocks and a comparison of *distant* events. Our assertion that the measurement of any velocity in one direction presupposes a knowledge of simultaneity is therefore correct.

Thus we are faced with a circular argument. To determine the simultaneity of distant events we need to know a velocity, and to measure a velocity we require knowledge of the simultaneity of distant events. The occurrence of this circularity proves that simultaneity is not a matter of knowledge, but of a coordinative definition, since the logical circle shows that a knowledge of simultaneity is impossible in principle.

We also notice that the second characteristic of a coordinative definition, namely its arbitrariness, is satisfied. It is arbitrary which time we ascribe to the arrival of the light ray at B. If we assume it to be 12:03, the velocity of light becomes equal in both directions. If we assume it to be 12:02, the light ray requires 2 minutes in one direction and 4 minutes in the other; this assumption is equally compatible with Fizeau's measurements. It does not make sense, therefore, to call the time 12:02 false or improbable, since we are here concerned not with an empirical statement but with a definition. This definition determines at once the velocity of light and simultaneity, and such a determination can therefore never lead to contradictions. If we wish to determine by velocity measurements which events are simultaneous, we shall always obtain that simultaneity which has already been introduced by definition.

It is this consideration that teaches us how to understand the definition of simultaneity given by Einstein

$$t_2 = t_1 + \tfrac{1}{2}(t_3 - t_1) \qquad (1)$$

which defines the time of arrival of the light ray at B as the mid-point between the time that the light was sent from A and the time that it returned to A. This definition is essential for the special theory of relativity, but it is not epistemologically necessary. Einstein's definition, too, is just one possible definition. If we were to follow an arbitrary rule restricted only to the form

$$(2) \qquad t_2 = t_1 + \epsilon(t_3 - t_1) \qquad 0 < \epsilon < 1 \qquad (3)$$

it would likewise be adequate and could not be called false. If the special theory of relativity prefers the first definition, i.e., sets ϵ equal to $\frac{1}{2}$, it does so on the ground that this definition leads to simpler relations. It is clear that we are dealing here merely with descriptive simplicity, the nature of which will be explained in § 27. The arbitrariness is restricted only by condition (3) which specifies that t_2 must lie between t_1 and t_3; otherwise the signal would arrive at B at a time earlier than its departure from A. The epistemological significance of this restriction will be discussed in detail in § 22.

These considerations have shown that simultaneity is a matter of a coordinative definition. Simultaneity also has the peculiar dual character which we can most easily observe in the definition of the unit of length. What we mean by a unit of length can be defined conceptually: a unit of length is a distance with which other distances are compared. Which distance serves as a unit for actual measurements can ultimately be given only by reference to some actual distance. The same is true of simultaneity. We can give a conceptual definition of "simultaneity": two events at distant places are simultaneous if the time scales at the respective places indicate the same time value for these events. What time points of parallel time scales do receive the same time value can ultimately be determined only by reference to actual events. This reference is essentially of the form: "These particular events are to be called simultaneous." We say with regard to the measuring rod as well as with regard to simultaneity that only "ultimately" the reference is to be conceived in this form, because we know that by means of the interposition of conceptual relations the reference may be rather remote. We may recall here the example of the determination of a unit of length by reference to a color, mentioned in § 4, where the reference is not directly to a spatial distance. Correspondingly, we find that the reference in the definition of simultaneity is commonly not in terms of the occurrence of arbitrary events, but in terms of light, i.e., a physical process, the properties of which are utilized in the definition of simultaneity. In this fashion we are able to replace a direct reference by a description of operations which can easily be repeated, since it is commonly understood what is meant by "light" and by these operations. The definition of simultaneity through the use of light signals, for instance Einstein's definition, cannot be compared to the definition of the meter by means of the Parisian standard meter, but is to be compared to the definition of the meter by means of the earth's circumference. In this definition the physical phenomenon *the earth's circumference* corresponds to the physical phenomenon *light* in the definition of simultaneity, and the rule "count off 40 million times" corresponds to the rule "send a light signal from A to B and back and set the time of arrival at B equal to the average of the two time values at A." Such a rule does not change the nature of the coordinative definition, since what is meant by "light" and "the circumference of the earth" can ultimately be determined only through a direct reference.

The conceptual definition which we related to the coordinative definition of simultaneity may appear empty; it is tautological to define simultaneity as the equality of time values on parallel time scales.

But the situation is no different for any other conceptual definition. All conceptual definitions are tautological in this sense, since they deal exclusively with analytic relations. A concept is coordinated to a combination of certain other concepts and derives its meaning only from these other concepts. The conceptual definition of the unit of length is also a tautology in this sense. Yet the desire for a different conceptual definition of simultaneity has a certain justification. We mean more when we speak of simultaneity; we are searching for a rule that restricts the determination of the parallel time scales in a special fashion. An answer to this question can only be given by the causal theory of time which we shall develop in § 21 and § 22. We anticipate, however, that this investigation will not eliminate the relativity of simultaneity but only justify the restriction of arbitrariness given in (3).

§ 20. ATTEMPTS TO DETERMINE ABSOLUTE SIMULTANEITY

Before we proceed from these results to further problems, we shall first discuss some of the objections that have been raised against the arbitrariness of simultaneity. The answers to these objections will assure us that the solution of the problem of simultaneity is correct. These criticisms consist in various attempts to establish absolute simultaneity.

The first of these attempts starts with the idea of using velocities greater than the velocity of light. As a result, the interval $t_3 - t_1$ of definition (2, § 19) would be shortened and the definition of simultaneity would become less arbitrary. If there existed a signal with infinite velocity, the interval would equal zero and absolute simultaneity would be established. Even if an infinite velocity could not be attained, the inaccuracy could be made as small as desired by means of correspondingly high velocities. Such an approximation would suffice to define absolute simultaneity as a limit. Indeed, if arbitrarily high velocities could be reached, there would be absolute simultaneity. The relation between signal velocity and the interpretation of the word "absolute" will be discussed in § 22. We may comment at this place, however, that this objection is pointless, since there are no signals that travel faster than light. We do not mean merely that physics has not yet discovered a higher velocity, but rather the positive assertion that there can be no higher velocity. Reasons for making this assertion will be given in § 32.

The second group of attempts to define absolute simultaneity uses specially conceived mechanisms. We can imagine an electrical mechanism of this type to be built according to Fig. 17.[1] The current

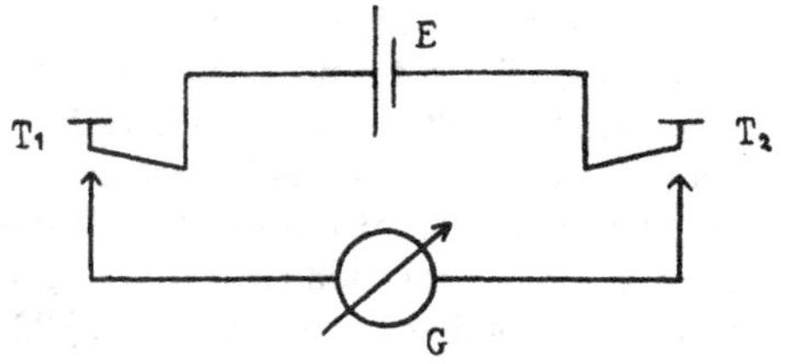

Fig. 17. An attempt to determine absolute simultaneity by means of an electrical arrangement.

from battery E flows through galvanometer G if the two switches T_1 and T_2 are closed. If only one of the two switches is closed, no current flows through G. Let us imagine now that T_1 and T_2 are closed for an instant. If these two instants are simultaneous, the circuit will be closed and a brief impulse of electric current will flow through G which will show a single deflection. If they are not simultaneous, the circuit will never be closed and no deflection will occur in G. The deflection in G is therefore the criterion for the simultaneity of the two events, and this simultaneity is defined without arbitrariness.

However ingenious this attempt may appear, closer examination reveals where it fails. It uses a far too primitive theory of electric currents. The property of an electric current to flow only in a completely closed circuit holds for stationary states alone. Under rapidly changing conditions, however, the electric current shows entirely different properties. The occurrences during a momentary closing of the two switches can be described as follows. Let us assume that only T_1 is closed for a moment. The electromagnetic field, which had previously spread from E to T_1, proceeds now to G (and further to the lower side of T_2). The electrons in the wire are set in motion and a short impulse of current flows through G. The characteristics of this current are those of a displacement current because of the capacity of the open contact T_2. G will thus show a deflection if it is a sensitive galvanometer, which, of course, must be assumed for these experiments. Thus we observe a deflection even though only one of the two switches

[1] This mechanism was described by F. Adler, in *Ortszeit, Systemzeit, Zonenzeit*, Vienna 1920, p. 81.

is closed. If in addition we close switch T_2, a second impulse of current results which increases the deflection in G. It is irrelevant here whether T_1 and T_2 are closed simultaneously or within a short interval of time. In either case we obtain the same deflection consisting of the sum of the two impulses.

Only if the difference in time is so great that the disturbance of the electromagnetic field which spreads from T_1 through G has already reached T_2 when the second switch is closed, will there be a difference in the magnitude of the deflection in G. The propagation of the electromagnetic field from T_1 to T_2 travels however with the velocity of light; thus there exists a small interval of time within which the two impulses of current may follow one another without any difference in the effect on G. It is therefore not permissible to conclude from the deflection in G that the two switches were closed simultaneously; they might as well have been closed within a small interval of time.

This mechanism therefore does not yield a decisive method for simultaneity. It leaves as much arbitrariness as the determination of simultaneity by means of signals, since the signal in this case is an electromagnetic disturbance which likewise propagates with the speed of light. The entire arrangement is really nothing but a disguised signaling process. What happens when the circuit is closed at T_2 depends, according to the law of action by contact, only on the state of the electric field in the immediate environment of T_2. Whether T_1 is open or closed is therefore irrelevant. Only if the disturbance of the field, caused by the earlier closing of T_1, has already advanced to T_2 will there be any effect on the happenings at T_2. In this case the circuit must have been closed at T_1 just early enough to permit the disturbance to travel the distance T_1GT_2 with the velocity of light. Letting this time interval be Δt, we can state that the magnitude of the deflection in G tells us only whether the difference in time between the closing of switches T_1 and T_2 is greater than Δt. If this difference is less than Δt, it is impossible to decide whether or not the two switches were closed simultaneously.

The electrical mechanisms for the determination of absolute simultaneity fail because electric effects propagate with the velocity of light. The relations of a stationary circuit suggest at first sight action at a distance, but actually no violation of the principle of action by contact occurs. The principle of action by contact is one of the most basic laws of physics. It is impossible for the effect of an occurrence to be immediately noticeable at any arbitrary distance. The effect spreads by traveling through all intermediate points. This principle applies to *every* form of causal propagation. Gravitation, for instance, offers no exception and the Newtonian law of gravitation is correct only under stationary conditions. For rapidly moving systems, this law must be corrected to account for the finite speed of the propagation of gravitation.[1] One should not conclude, however, that the principle of action by contact necessarily implies the existence of a finite limit to the speed of all causal propagation. The principle excludes only infinite velocities, whereas it is compatible with any arbitrarily high velocity. If we wish to assert the existence of a finite limit, this assertion will have to be added to the principle of action by contact. Only this addition enables us to assert that there is no mechanism capable of determining absolute simultaneity. Even gravitational forces cannot be used for such a mechanism, since they also spread with the velocity of light.

We may therefore omit the different mechanisms that have been invented for the determination of absolute simultaneity, because all of them are variations of the same fundamental idea and fail because each presupposes, in a more or less disguised form, an infinite or arbitrarily high velocity of causal propagation. We shall mention one more example, which involves a misconception of rigid bodies. Let a rigid rod rest with its ends on the marks A and B (Fig. 18). If the rod

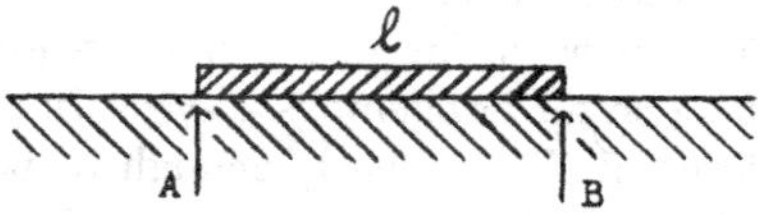

Fig. 18. An attempt to determine absolute simultaneity by means of moving a rigid rod.

is suddenly grasped at A and pulled to the left, its ends no longer cover the points A and B. The two moments at which the coincidences of A and B with the two ends of the rod are lost, must be absolutely simultaneous according to the laws of classical physics. This cannot be admitted in a physics which includes the principle of action by contact; in such a physics there are no absolutely rigid bodies. When the rod is grasped at A, the end which rested on B does not move

[1] This requirement is actually carried out in Einstein's theory of gravitation. At the same time, however, the Newtonian laws are recognized as approximations in yet another sense.

immediately, since the effect spreads by means of an elastic propagation from A to B. The velocity of this elastic propagation cannot be greater than the velocity of light (actually it is smaller). Therefore, this arrangement cannot be used for the determination of simultaneity. With the existence of a limit for all causal propagation, not only infinite velocities but absolutely rigid bodies are excluded.

Since absolute simultaneity cannot be attained through causal propagation, the possibility of a fundamentally different method might be envisaged. This leads to a third attempt at a determination of absolute simultaneity, to "absolute transport time."

This procedure uses the transport of clocks to establish absolute simultaneity. The two clocks are synchronized when close together (i.e., the hands are in the same position at the same time) and then one of them is moved. We then have a clock at a distant point synchronized with the first one. We shall call this arrangement of clocks a *transport-synchronization*.

A criticism of this method can be carried out in either of two ways. First we may investigate whether the transport of clocks actually leads to a simultaneity that is free from contradiction. This investigation presupposes that the time indications of clocks are independent of the path and velocity of transport, i.e., the following assertion would have to be true: two clocks, synchronized at one place, are still synchronized when they are brought to a different place along different paths with different speeds. This statement, however, is denied by relativistic physics (cf. § 30). Either alternative seems possible, but only experience can decide which of them holds for reality. Even without the assumptions of relativistic physics we can state that the possibility of transport-synchronization depends on an empirical assumption that must be tested.

However, if relativistic physics were wrong, and the transport of clocks could be shown to be independent of path and velocity, this type of time comparison could not change our epistemological results, since the transport of clocks can again offer nothing but a *definition* of simultaneity. Even if the two clocks correspond when they are again brought together, how can we know whether or not both have changed in the meantime? This question is as undecidable as the question of the comparison of length of rigid rods. Again, a solution can be given only if the comparison of time is recognized as a definition. If there exists a unique transport-synchronization, it is still merely a *definition* of simultaneity.

We can characterize the peculiarity of the transport-synchronization clearly if we use the Minkowskian picture of the four-dimensional space-time manifold. The transport of a clock constitutes a causal chain spreading from the point event E_1 to the point event E'_2. In this respect there is no difference between this method and the method which employs signals for the determination of simultaneity. Here too we find that the comparison of time is established with the aid of a causal chain. While the signal simultaneity uses the velocity of the causal propagation, the transport of clocks makes use of different considerations.

Here we can imagine points marked off on the causal chain $E_1E'_2$ (Fig. 19), which are produced by the transported clock whenever it

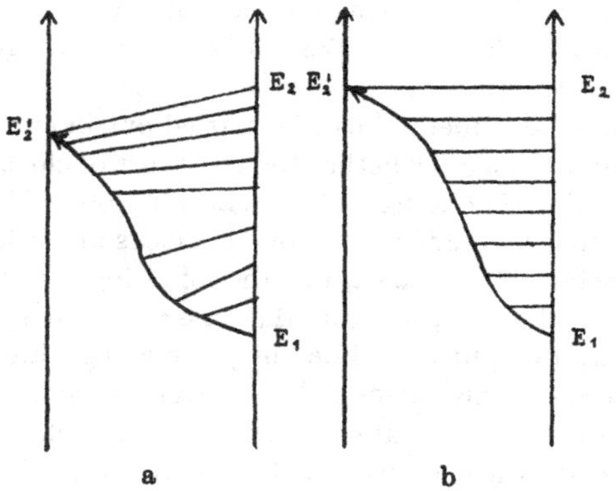

Fig. 19. Definition of simultaneity by means of the transport of clocks.

completes a unit of time. At the same time the clock that remains subdivides the chain E_1E_2, and the time comparison between E'_2 and E_2 results from counting the number of sections on the two causal curves. It is obvious that this comparison of time presupposes that the marked-off sections are of equal length. But this presupposition is based on a coordinative definition and the analogy to the definition of spatial congruence is clearly seen. The theory of absolute time states that the diagram necessarily looks like Fig. 19*b*, but it is clear that at first the clocks furnish only the irregular relations of Fig. 19*a* and that this figure must be *redefined* in order to coincide with Fig. 19*b*. An additional assumption, which alone makes such a redefinition univocal, must be recognized. It is the assumption that a clock which moves along a differently winding world-line from E_1 to E'_2 marks off the same number of sections. Of course, this is an empirical assumption.

The absolute transport time, if uniquely defined, would give us nothing but a *definition* of simultaneity, which is a definition in the same sense as the definition of congruence by means of rods. The theory of relativity, however, maintains the existence of an essential difference. Whereas the congruence of rods is independent of the path of transport, that of clocks is not. The theory of relativity excludes the transport time because of this physical fact.

We summarize the results of the preceding sections as follows. The time metric depends on three coordinative definitions. The first deals with the *unit of time* and determines the numerical value of a time interval. The second deals with *uniformity* and refers to the comparison of successive time intervals. The third deals with *simultaneity* and is concerned with the comparison of time intervals which are parallel to each other at different points in space. These three definitions are required in order to make a time measurement possible; without them the problem of the measurement of time is logically undetermined.

There is neither absolute simultaneity nor absolute uniformity, if we understand by "absolute" the property that this time is the only correct time. However, there remains the possibility that physical mechanisms or the entire system of physical laws might distinguish one definition as simpler than the others. In this sense there might be an absolute time. For instance, we know from experience that the definition of uniformity by means of clocks or of the law of inertia is distinguished from others by its simplicity. This distinction is maintained in the special theory of relativity and vanishes only for more general gravitational fields. Among the definitions of simultaneity, those based on the infinite limiting velocity or the transport of clocks might turn out to be the simplest. Whether or not they do is an empirical question; both possibilities are denied by the special theory of relativity. Hence this theory played an important role in the clarification of the definitional character of simultaneity.

§ 21. TIME ORDER

In the preceding sections we have developed the three *metrical* coordinative definitions of time, which concern the unit of time, uniformity and simultaneity. We shall now turn to another type of coordinative definitions, namely, the *topological* ones. We shall present two such definitions.

The first topological determination of time deals with *time order at the same point.* Consider an observer at a certain point who has to decide the temporal order of two events, both of which occur at the same place. This aspect of temporal order corresponds to the relation *greater than* in the series of numbers and we shall represent it by the same symbol. "$E_2 > E_1$" means therefore: "E_2 is later than E_1."

With respect to two events that are sufficiently separated in time, the observer has an immediate experience of time order, and he uses this experience as the basis for the ordering of the events. However, in this chapter, we shall not refer to the subjective experience of time order. Subsequently it will be shown that it is in principle impossible to use subjective feelings for the determination of the order of external events. We must therefore establish a different criterion.

Such a criterion is found in the causal relation. *If E_2 is the effect of E_1, then E_2 is called later than E_1.* This is the topological coordinative definition of time order. To complete this statement we should add that it also applies to the case where E_1 is only a partial cause of E_2 or where E_2 is only a partial effect of E_1.

It is obvious that with the definition of "later than" we have also given the definition of "earlier than." The second relation is nothing but the converse of the first. If E_2 is later than E_1, then E_1 is earlier than E_2. This follows analytically and needs no new coordinative definition.

However, we must now make sure that our definition of "later than" does not involve circular reasoning. Can we actually recognize what is a cause and what is an effect without knowing their temporal order? Should we not argue, rather, that of two causally connected events the effect is the later one?

This objection proceeds from the assumption that causality indicates a connection between two events, but does not assign a direction to them. This assumption, however, is erroneous. Causality establishes not a symmetrical but an asymmetrical relation between two events. If we represent the cause-effect relation by the symbol C, the two cases

$$C(E_1, E_2) \quad \text{and} \quad C(E_2, E_1)$$

can be distinguished; experience tells us which of the two cases actually occurs. We can state this distinction as follows:

If E_1 is the cause of E_2, then a small variation (a mark) *in E_1 is associated with a small variation in E_2, whereas small variations in E_2 are not associated with variations in E_1.*

If we wish to express even more clearly that this formulation does not contain the concept of temporal order, we can express it in the following form, where the events that show a slight variation are designated by E^*:

We observe only the combinations

$$E_1E_2 \qquad E_1{}^*E_2{}^* \qquad E_1E_2{}^* \tag{1}$$

and never the combination

$$E_1{}^*E_2 \tag{2}$$

In this arrangement the two events are asymmetrical and therefore it defines an *order*. That event which appears in the unobserved combination without an asterisk, namely E_2, is called the *effect* and furthermore the *temporally later* event.

It should be noted that assertions (1) and (2) were obtained without the presupposition of an order. We could have placed the event E_2 first in these combinations and would still have been able to distinguish E_2 as the effect. On the other hand, we have made the assumption that we are able to distinguish between E and E^*, i.e., that we know which of the two events has the special mark. We may do so because E^* is to be interpreted as a combination (E, e) where e signifies an additional event, namely, the special mark. Just as we may assume that E_1 and E_2 can be recognized as two separate events, we may assume that E^* can be recognized as the combination of two separate events. This assumption lies within the frame of our schematization, which presupposes that we can distinguish individual events. We cannot justify this schematization within the frame of our present discussion, which is limited to space and time, since such a justification belongs to the analysis of the concept of causality. Here we must be satisfied with the assertion that the space-time problem cannot be solved at all without some schematization.[1]

An example: We send a light ray from A to B. If we hold a red glass in the path of the light at A, the light will also be red at B. If

[1] A presentation of causal order which does not use the principle of the mark, but which of course presupposes certain schematizations, was given by the author in "Die Kausalstruktur der Welt und der Unterschied von Vergangenheit und Zukunft," *Berichte der Bayrischen Akademie*, math. naturwiss. Abh., 1925, p. 133. This more rigorous presentation is not possible, however, without an introduction of the concept of probability. Reference is also made to a remark in my book *Axiomatik der relativistischen Raum-Zeit-Lehre*, Braunschweig, 1924, p. 133, which refers to a possible connection of the mark principle with the second law of thermodynamics.

we hold the red glass in the path of the light at B, it will *not* be colored at A.

Another example: We throw a stone from A to B. If we mark the stone with a piece of chalk at A, it will carry the same mark when it arrives at B (event E_2). If we mark the stone only on its arrival at B, then the stone leaving A (event E_1) has no mark.

This distinction appears trivial, but it is extremely significant. A theory of causality which ignores this elementary difference has neglected the most essential aspect. The procedure which we have described is used constantly in everyday life to establish a time order, and we have no other method in many scientific investigations where time intervals are too short to be directly observable. We must therefore include the mark principle in the foundations of the theory of time.

We have in the above principle a criterion for causal order that does not employ the direction of time, and we can therefore use it in our definition of time order. There exists a topological coordinative definition for time order. We can base it in general on the concept of the *causal chain*, in which the order of events corresponds to the order of time. Occasionally one speaks also of *signals* or *signal chains*. It should be noted that the word "signal" means the transmission of signs and hence concerns the very principle of causal order which we have discussed.

We have to distinguish here between two problems. First, the procedure described leads to an *order* of time, in the same sense in which the points on a line are ordered. Such a series of points has two directions, neither of which has any distinguishing characteristic. Temporal order, too, has two directions, the direction to earlier and the direction to later events, but in this case one of the two directions has a distinguishing characteristic: time flows from the earlier to the later event. Time therefore represents not only an ordered series generated by an asymmetrical relation, but is also *unidirectional*. This fact is usually ignored. We often say simply: the direction from earlier to later events, from cause to effect, is the direction of the progress of time. However, in this form the assertion is empty unless we specify what "progress of time" means. In the same fashion we could say that the points on a line progress from left to right; but this assertion is empty, since the progress of points means here nothing but the progressing in the selected direction. When we speak about the progress of time, in contrast, we intend to make a synthetic assertion which refers both to an immediate experience and to physical reality. This particular problem can only be solved if we can formulate the content of the assertion more precisely. We shall leave this problem for the time being and content ourselves with the conclusion that the direction which we have defined as earlier-later is the same direction as that of the progress of time. For the problems dealt with in the theory of relativity it suffices that there exists a serial *order* of time, i.e., that we can distinguish between two directions which are opposite to each other. At present, we shall not refer to the unidirectional character of time.

We must now investigate the question whether the time order defined above can be carried through consistently.

If a contradiction did occur, it would manifest itself as follows: According to one analysis, E_2 would be later than E_1, and according to another E_2 would be earlier than E_1. This result would lead to the schema of Fig. 20, in which there is a causal chain from E_1 to E_2 such

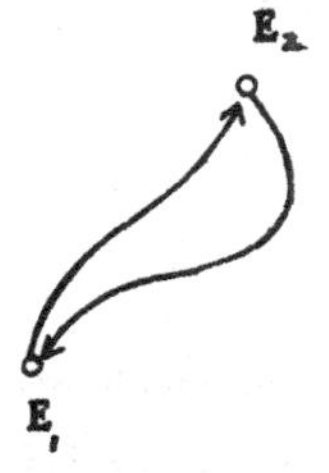

Fig. 20. A closed causal chain.

that E_2 is an effect of E_1, while there is also a causal chain that makes E_1 an effect of E_2. The combination would result in a causal chain that returns to its origin. In order to exclude such a contradiction we must make the assumption that *there are no closed causal chains.*

At first glance, one might suppose that this assumption goes too far and that indeed there occur closed causal chains in certain mechanisms, the prototype of which is the electric bell. The pulling (P) of the lever causes a break (B) in the current, which in turn causes the return (R) of the lever; this switches on (S), the current which finally again pulls (P) the lever. It appears as though this chain of events could be diagramed as in Fig. 21a, which represents a closed curve. The mistake in this argument, however, is easily seen. The individual pulls of the lever are different events; i.e., although they are of the same kind, they are not identical events. Consequently, the chain should be diagramed as in Fig. 21b, namely, as an open chain. Our principle of the mark forces us to this conception. If we make some change in B_1, for instance by short-circuiting the current and thus preventing the return of the lever, then P_2 will no longer occur. However, such interference does not change P_1. Now it is clear what we mean by a closed chain, namely, a chain that returns to *identically the same event*, not to one of the same kind.

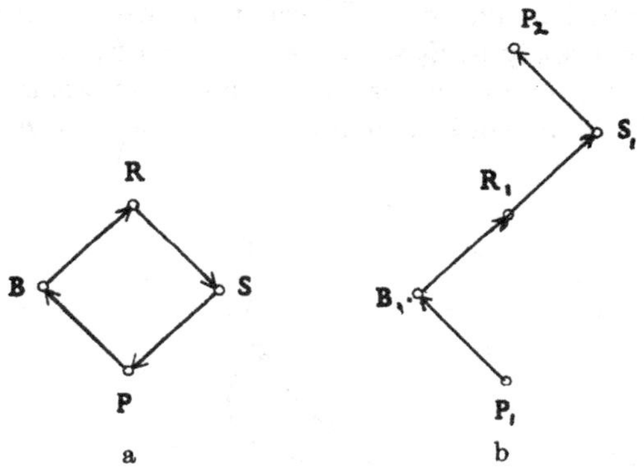

Fig. 21 *a* and *b*. The mechanism of an electric bell.

It seems obvious that there are no such chains. Yet the statement that there are no closed causal chains expresses merely a conclusion drawn from experience, which has thus far been confirmed without exception. We could imagine experiences which would disprove this statement.

Such conceptions are not unfamiliar. In the form of a periodic return of all physical events, they have played a role in many cosmologies. It is conceivable—although of course there is no evidence for it—that some day the entire universe will return to a previous state in every detail and start from there anew on the identical course of events. Such an occurrence would leave us the choice of interpreting the chain of events either according to Fig. 21a or according to Fig. 21b. As in the problem of space we have here the choice of regarding two states of the same kind either as *identical* or only as *similar*. Our principle of the mark fails here for a reason to be explained later. This property of the world as a whole has no significance for individual events and we can thus limit the definition of time order for our purposes to causal chains within one world period, the starting point of which is to be considered arbitrary. It is therefore not the case of a periodic universe, but a more complicated one, which interests us as a counterexample to our assumption.

Let us assume that some *individual* world-lines are closed whereas others are not. We shall examine this assumption by the help of an example, in order to see what fundamental principles would have to be abandoned.

World-lines I and II of Fig. 22 are both world-lines of human beings. World-line I is normal, while world-line II does not intersect itself

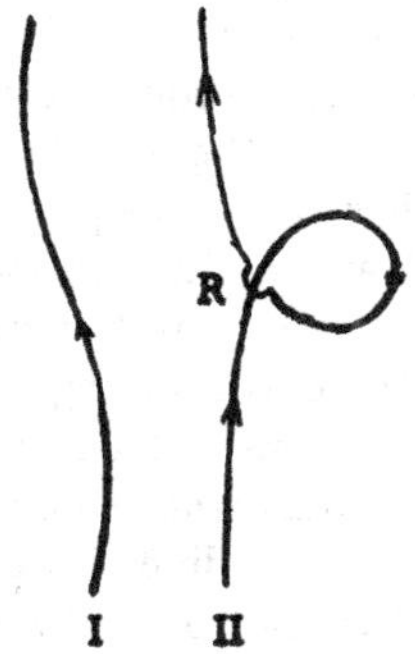

Fig. 22. A closed and an open world-line next to each other.

directly, but is represented by a curve which, like a spiral, is not really closed but merely returns to the neighborhood of one of its points. This fact is indicated by the little arc at that particular point. A causal connection between the neighboring parts of this world-line can be established by means of signals (speech) within the region of the small arc. If you were the individual of world-line II, you would have the following experience.

Some day you meet a man who claims that you are his earlier self. He can give you complete information about your present condition and might even tell you precisely what you are thinking. He also predicts your distant future, in which you will some day be in his position and meet your earlier self. Of course you would think the man insane and would walk on. Your companion on world-line I agrees with you. The stranger goes his way with a knowing smile; you lose sight of him as well as of your companion on world-line I and forget about both of them. Years later you meet a younger man whom you suddenly recognize as your earlier self. You tell him verbatim what the older man had told you; he doesn't believe you and thinks you are insane. This time you are the one that leaves with a knowing smile. You also see your former companion again, exactly as old as he was when you last saw him. However, he denies any acquaintance with you and agrees with your younger self that you must be insane. After this encounter, however, you walk along with him. Your younger self disappears from sight and from then on you lead a normal life.

These events would be very strange, but not logically impossible. We can now recognize those fundamental principles which would have to be abandoned if these events should actually occur.

We could no longer speak of the uniqueness of the present moment. On the same world-line there would be periodic "now-points" one after the other. In region R we would find two now-points of the same world-line in causal interaction; and under these circumstances we would lose the possibility of conceiving of the self as one identical individual in the course of time. There would be on this world-line a succession of new individuals who would travel the same world-line at certain intervals. On world-line I we must also mark off such periods; however, the individuals of this world-line would never notice one another since their now-points never enter into causal interaction.

We can thus recognize that not only the uniqueness of time order, but also the identity of the individual during the passage of time, would be lost. This is the main difficulty in trying to imagine such events. We also recognize those properties of the causal chain which underlie the familiar concept of individuality. This concept originates in the fact that there are no closed causal chains.

We face here a fundamental principle controlling physical reality. It enables us to speak of a unique time order and of a unique now-point. Furthermore, it makes possible the concept of the individual that remains identical during the passage of time.[1] It is therefore the most

[1] This is often denoted by the term *genidentity*. The term was introduced by K. Lewin, in *Der Begriff der Genese*, Berlin 1922.

important axiom regarding time order, and we realize to what an extent the familiar concept of time order is based on this characteristic of causality. Of course, this axiom is a result of experience; hence events of the type described above cannot be excluded *a priori.*

Besides this axiom there are several other less important axioms regarding time order. They deal with the continuity of time order and are likewise empirical statements about the nature of causal relations.

§ 22. THE COMPARISON OF TIME

We now turn to the comparison of two time series at different points in space. For this purpose we shall again use signals. We must first make the assumption that there always are connecting signals between any two points in space. Since we use signals, our previous definition of time order offers an important result. Let E be the event of the departure of the signal from P, and E' the event of its arrival at P'; then E and E' are two events connected by a signal; consequently they are ordered; E' must be later than E. Certain events are therefore already ordered although they belong to different temporal sequences.

Not all events are ordered, however. To clarify this situation, we shall introduce the auxiliary concept of *first-signal.* If several kinds of signals are sent from P at the same moment, they will arrive at P' at different times. To order the times of their arrivals at P' we need only the time series at P'. That particular signal having the earliest time of arrival at P' is called the first-signal; it is therefore defined as the fastest message carrier between any two points in space. The existence of first-signals can be derived from our previous axioms.

We now send a first-signal from P, calling the event of its departure E_1 (Fig. 23). The event of its arrival at P' is called E'. Simultaneously with the arrival of this signal, another first-signal is sent from P'. The arrival of this signal at P is the event E_2. We ask for the order of E' relative to E_1 and E_2. According to definition we have

$$E' \text{ is later than } E_1$$
$$E' \text{ is earlier than } E_2$$

Let E be an event at P between E_1 and E_2. What is the position of E relative to E'? Here, our definition of time order fails. A first-signal sent from E would arrive at P' later than E', and a first-signal from E' arrives at P later than E, as can easily be seen from Fig. 23.

It is therefore impossible to connect the events E and E' in either direction by a signal, and their time order is consequently not determined. We shall call such events *indeterminate as to time order.*

We must now distinguish between two cases. In the case of Fig. 23, the time interval E_1E_2 is coordinated to the event E', and every event of this time interval except for the end-points is indeterminate

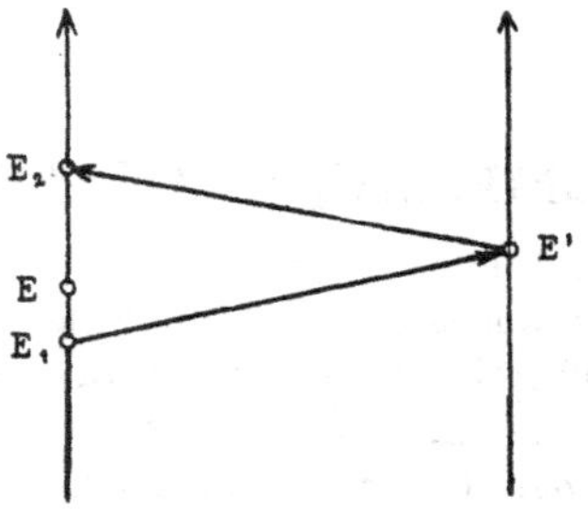

Fig. 23. Events which are indeterminate as to time order.

as to time order relative to E'. The other case would occur if there were no limit to the speed of signals; then the first-signal would have an infinite velocity [1] and the time interval E_1E_2 would be reduced to a single point E_0. This point E_0 would then be coordinated to E' and be the only event at P indeterminate as to time order relative to E'.

It is an empirical question which of these two cases occurs in our world. According to classical physics it is the second case, according to relativistic physics the first. There is decisive empirical evidence for the relativistic theory of time. Light has the limiting velocity; it is physically impossible to reach higher velocities. *Light is therefore a first-signal.* For the experimental basis of this statement see § 32. It is important to note that we are able to formulate the limiting character of the velocity of light without the concepts either of velocity or of simultaneity: the departure and return of the first-signal $PP'P$ are separated at P by a finite interval of time.

This result leads to a clarification of the problem of simultaneity. The definition of simultaneity ascribes equal time values to different

[1] This means that the first signal itself would not be a signal, but the limit of all signals. Consequently, two events connected by a first-signal would not yet be temporally ordered. This justifies our subsequent assertion that in this case E_0 would be indeterminate as to time order relative to E'.

points in space. It must not contradict our definition of time order, which restricts the time values of E_1, E', and E_2 in the sense of (3); therefore, only events which are indeterminate as to their time order may be regarded as simultaneous. Among the events, however, no further rule restricts our choice. We define: *any two events which are indeterminate as to their time order may be called simultaneous.*

This topological definition would be sufficient for a unique definition of simultaneity in the classical theory of time. To each event E' at P' there corresponds a *single* event E at P which is indeterminate as to time order, and this E would then be regarded as simultaneous with E'. In the case of a finite limiting velocity as in Fig. 23, the topological definition of simultaneity does not lead to a unique determination, but admits any event E between E_1 and E_2 as simultaneous with E'. The relativity of simultaneity follows from the topological peculiarity of the causal structure, according to which first-signals (Fig. 23) determine *a finite interval* and not a *point-event,* as corresponding to a single event E' located on a different space point.

These considerations supply the conceptual definition of simultaneity which we previously sought (p. 129). The concept *simultaneous* is to be reduced to the concept *indeterminate as to time order.* This result supports our intuitive understanding of the concept simultaneous. Two simultaneous events are so situated that a causal chain cannot travel from one to the other in either direction. Events which occur at this moment in a distant land can no longer be influenced by us, not even by telegram; and conversely, they can have no effect on what is happening here at the present moment. *Simultaneity means the exclusion of causal connection.* Within the frame of an epistemological investigation this result appears justified. Yet we must not commit the mistake of attempting to derive from it the conclusion that this definition coordinates to any given event a single event at a given different place. This would be the case only for a special form of causal structure, a form that does not correspond to physical reality. The causal structure of our universe involves the consequence that exclusion of causal connection does not lead univocally to a unique simultaneity.

Our epistemological analysis thus leads to the discovery that the relativity of simultaneity is compatible with the intuitive conception which we connect with simultaneity. It is not this conception which is incorrect, but the conclusion derived from it that simultaneity must be uniquely determined. Thus all the difficulties which philosophers have seen in the relativity of simultaneity are eliminated. At the same time our result eliminates the mystery with which adherents of the theory of relativity have invested this concept. There are essentially two errors in relativistic presentations which have confused the epistemological issues.

One mistake results from the derivation of the relativity of simultaneity from the different states of motion of various observers. It is true that one *can* define simultaneity differently for different moving systems, which incidentally is the reason for the simple measuring relations of the Lorentz transformation, but such a definition is not *necessary*. We could arrange the definition of simultaneity of a system K in such a manner that it leads to the same results as that of another system K' which is in motion relative to K; in K, ϵ would not be equal to $\frac{1}{2}$ in the definition of simultaneity (§ 19), but would have some other value. It is a serious mistake to believe that if the state of motion is taken into consideration, the relativity of simultaneity is necessary. Actually the relativity of simultaneity has nothing to do with the relativity of motion. It rests solely on the existence of a finite limiting velocity for causal propagation. The arbitrariness in the choice of simultaneity makes it possible to assign the value $\epsilon = \frac{1}{2}$ to *every* uniformly moving system (or inertial system). A further confusion arises if in addition to the relativity of motion, the subjectivity of the observer is introduced into the argument. The relativity of simultaneity has nothing to do with the subjectivity of sense perception. The visualization of several logically equivalent methods of measurement is merely facilitated when different definitions of simultaneity are ascribed to different observers. The multiplicity of observers in the theory of relativity has no further significance. We are here concerned not with a difference of reference points but with a difference in the logical presuppositions concerning measurements. The comparison of time must be defined in some fashion before time measurements are possible at all. That this determination is arbitrary within certain limits, is due to the causal structure of the world and rests therefore on empirical grounds.

The other error committed in certain relativistic presentations lies in the belief that an absolute theory of time is a logically impossible conception. Such a criticism applies only to the conceptions of absolute time usually presented by the opponents of the theory of relativity. The following formulation, however, defines a meaningful concept of absolute time: absolute time would exist in a causal structure for which the concept *indeterminate as to time order* lea simultaneity, i.e., for which there is no finite interval of the departure and return of a first-signal $PP'P$ at P precise formulation reveals the error in the classical the ...e: this property of the causal structure was *postulated a priori*, when an *empirical* investigation was called for. Relativistic physicists have indeed formulated a correct theory of time, but they have left their opponents in the dark concerning the epistemological grounds of their assumptions.

§ 23. UNREAL SEQUENCES

Let us consider one more objection against the theory of time that we have developed. We have based the relativity of simultaneity on the finiteness of the velocity of light. Is it not possible, however, to produce arbitrarily high velocities even if we recognize the limiting character of the velocity of light?

Consider, for instance, a mechanism of the following kind (Fig. 24). Two rulers which cross obliquely lie one above the other. If the

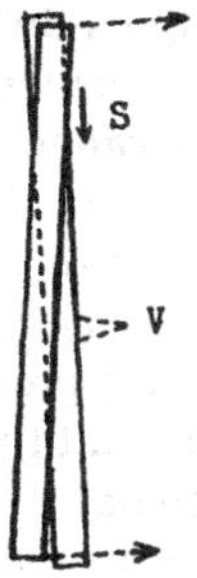

Fig. 24. The moving point of intersection of two rulers.

obliquely drawn ruler moves in the direction of the arrow, the point of intersection S moves downward along the edge. The smaller the angle between the rulers, the greater will be its velocity. If they are exactly parallel, the velocity of the point of intersection becomes infinite. Can we not consider this point of intersection as a moving signal and establish thus the infinite velocity which we need for the definition of absolute simultaneity?

This question can be answered when we consider our definition of a signal. The point of intersection cannot transfer a mark. If we add, for instance, to the lower ruler a projection V (indicated by dots), the signal would be interrupted. On the other hand, the signal arrives unchanged in the lower part of the ruler. The "signal" consisting of the moving point is therefore not a real process, and not a signal, strictly speaking; we shall call it an *unreal sequence*. Because of its properties it does not define a direction. If we call the departure of the point of intersection from the upper end E_1 and its arrival at the lower end E_2, we have to order the four combinations (1 and 2, § 21). We observe

$$E_1E_2 \qquad E_1{}^*E_2 \qquad E_1E_2{}^* \tag{1}$$

but never

$$E_1{}^*E_2{}^* \tag{2}$$

In contrast to the order of (1) and (2) in § 21, we find here that E_1 and E_2 appear *symmetrically*. Consequently, it cannot be determined on the basis of this phenomenon in which direction the point of intersection travels.

It may be objected that the direction of motion of the point of intersection can be recognized in a different manner: it depends on the direction in which the ruler is slanted. This objection overlooks the fact that the slope of the ruler cannot be determined directly since we are dealing with a ruler *in motion*. Its slope can be defined only in the sense of the direction of a moving line; for this purpose we project every point of the ruler simultaneously on a system at rest and measure the slope of the projection. This procedure will be discussed in detail in § 25. Here we shall say only that the result depends on the definition of simultaneity. According to one definition the lines are parallel and the point of intersection will move infinitely fast; according to another it will move upward; and according to a third it will move down. The motion of the point of intersection therefore cannot be used for a definition of simultaneity. On the contrary, the determination of the direction of the motion of the point of intersection demands a prior definition of simultaneity.

The same can be demonstrated for all unreal sequences, many of which can be constructed. Another example would be the lateral motion of a light-ray sent out by a quickly rotating searchlight which at a certain distance has a speed higher than that of light.

Although the direction of rotation of the lamp cannot be inverted because the lamp itself is a physical mechanism, we can define simultaneity such that outside a certain circle the light travels in a direction opposite to that of the lamp. For a proof we must refer to a different publication.[1]

The discussion of unreal sequences shows clearly the significance of the causal theory of time. Only the events in a real causal process are temporally ordered; unreal sequences obtain their time order as the result of some method of time comparison which is already defined within the system. With this distinction, the relativistic theory of time elucidates the phenomenon of time better than the classical theory.

[1] A., § 26.

CHAPTER III. SPACE AND TIME

A. The Space-Time Manifold without Gravitational Fields

§ 24. THE PROBLEM OF A COMBINED THEORY OF SPACE AND TIME

So far we have treated space and time separately. We have described the specific problems that presented themselves in each of the two types of order, and have developed the epistemological basis on which we can now construct a combined theory of space and time. Such a construction must be the final goal of any epistemological investigation of space and time, since physical events are ordered in space as well as in time. Hence it is only this combined order which presents the final solution of the space-time problem. Such a solution is difficult, because the *combination* of the two orders introduces specific problems that do not appear in the study of either of them alone. This fact justifies the treatment of these problems in a separate chapter on the combined space-time order.

The present investigation will lead us deeper into physics than was necessary in the previous two chapters. We now want to develop the actual construction of the space-time metric as well as the epistemological principles on which the space-time theory is based. This presentation will lead us a step beyond the epistemological framework of this book, since such a construction is actually the task of physics, which uses the metric constantly. The step will appear advisable when one considers how little attention physicists have paid to the epistemological aspects of this problem. Presentations of the space-time theory by physicists are concerned primarily with physical issues and fail to make the epistemological problems explicit. On the contrary, they are often epistemologically vague, in order to make the physical theory appear as plausible as possible. In this chapter we shall present the physical theory of space and time including certain aspects of the theory of gravitation. We may defend this procedure on the grounds that the drawing of sharp boundaries between the various sciences is detrimental to their epistemological clarification. A philosopher constantly afraid of stepping into another domain runs the danger of asserting empty generalities, because his philosophy is not sufficiently anchored in the specific sciences. It would be rewarding if the present epistemological study should also help to clarify some problems for the physicist.

We shall devote Part A to problems connected with the special theory of relativity. We shall, however, treat only the theory of space and time developed by the theory of relativity, omitting related problems, e.g., of electrodynamics, which fall outside the frame of our investigation. We shall always keep epistemological interests in the foreground; yet we need to show what physical considerations pertaining to space and time are presented in the special theory of relativity, and how these considerations provide physical content for the epistemological frame developed in the foregoing chapters.

The restriction to gravitation-free spaces, necessary for the special theory of relativity, finds its historical origin in the Newtonian principle of relativity which introduced the uniformly moving *inertial system* as the normal system. For Newton the problems of gravitation begin with acceleration. This far-reaching idea was adopted by Einstein with certain reservations, to the extent that uniformly moving systems occupy a central position in his theory. He carried the Newtonian relativity a step further with his principle of relativity, which excludes axiomatically the possibility of assigning a preferred position to any single system within this class. According to Newton it is only within *mechanics* that no distinction between inertial systems can be made. When Newton believed, in addition, that even the motion of light does not furnish a distinction, he could do so only because he considered light as the emission of small particles, i.e., as a mechanical process. But in view of further developments in optics which led to the wave theory of light and with it to the ether as medium of the waves, one of the inertial systems must be singled out by the motion of light as the system at rest relative to the ether.

Since Einstein did not draw this conclusion in spite of the wave theory of light, he has extended the Newtonian relativity to the field of optics.

Today even this form of relativity theory has become a special case in the framework of Einstein's general theory of relativity, which includes the phenomena of gravitation. According to this theory a *metrical* field pervades the entire space and in general does not satisfy the conditions of the special theory of relativity. Only in special cases, for instance in the vast empty spaces between the fixed stars, does the metrical inertial field permit the construction of inertial systems of greater extension. This far-reaching consideration leads to a complete clarification of the special theory of relativity and explains what determines the unique state of motion of the inertial systems. We shall present these ideas in Part B and content ourselves in Part A with the axiomatic supposition that there are spaces, or parts of space, to which the special theory of relativity can be applied, and that in these spaces a certain state of motion can be distinguished as uniform. It will be our task to show how this particular state of motion can be recognized and determined by certain physical phenomena. This approach is advisable because, as we shall show in Part B, it can be extended to spaces that contain gravitation provided that we restrict ourselves to infinitesimal regions, and because it supplies in this form the foundation also for the general theory of relativity. The state of motion, as well as a detailed construction of the space-time metric, will be the subject matter of our presentation.

We must begin this analysis, however, with a preliminary investigation. In the preceding chapter we developed the relativity of simultaneity; and before we begin with the metrical construction, we must clarify the consequences of this relativity for the measurement of space. It is here that we shall find the new ideas which the epistemological analysis of the concept of time has introduced into the combined theory of space and time.

§ 25. THE DEPENDENCE OF SPATIAL MEASUREMENT ON THE DEFINITION OF SIMULTANEITY

The fact that the definition of simultaneity can be arbitrarily chosen leads to consequences for the measurement of space which become apparent when systems *moving with different velocities* are considered.

Our previous conception of the measurement of spatial distances was based on the transport of measuring rods; the value obtained by successively marking off the measuring rod along the segment is its length. However, this definition of length is applicable only if the measuring rod is *at rest* relative to the segment. Although the measuring rod is moved when it is placed repeatedly along the segment, it is at rest relative to the segment at the particular moment when it is marked off, i.e., at the moment when it fulfills its metrical function. The length obtained in this fashion is commonly called the *rest-length*; it is the length of a segment measured with a measuring rod relative to which it is at rest.

First we find that there exists an ambiguity in the comparison of the *rest-length of measuring rods moved with different velocities*. If we carry a measuring rod into a system in a different state of motion, it is impossible to compare it in the described manner with a measuring rod at rest; therefore, a coordinative definition must be introduced. We say that the measuring rod is to be regarded as having the same length whether at rest or in motion. This determination has nothing to do with simultaneity; and we shall call it the *first comparison of length in kinematics*.

We must next lay down a rule which we shall call the *second comparison of length in kinematics*, and which deals with the question of the *length of a moving line-segment*. This problem appears when all lengths are referred to a single coordinate system K (Fig. 25), which may

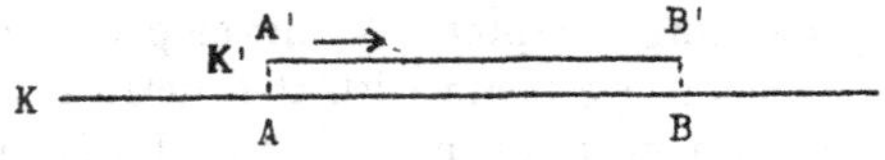

Fig. 25. The length of a moving line-segment.

be conceived as a framework of rigid rods. We may then perform measurements only with rods which are at rest relative to the coordinate system. How can we determine, under these circumstances, the length of a segment $A'B'$ which is moving in the direction of the arrow? If we were to mark it off with the measuring rod, this would give us the rest-length in K', not its length as measured in K.

This measurement must therefore be made by some indirect method which includes time measurements. Let us suppose that a definition of simultaneity has been given for K and that at a given instant, defined for K, we mark on the axis of K those points which coincide with the endpoints A' and B' of the moving segment. The moving segment is thus *projected* on the system K and the length of the projection AB can now be determined by a measuring rod which is at rest in K. We shall call the rest-length of AB the *length of the moving segment $A'B'$ measured in K.*

How do we know that this method is correct? This question is meaningless, since we are again concerned with a definition. What is meant by the length of a moving segment? In classical kinematics it was never noticed that such a concept had to be introduced, because the usual method determines only the rest-length of a segment. The length of a moving segment was tacitly assumed to be identical with its length at rest. We can see, however, that in this manner only the rest-length of a moving segment is defined analogous to the rest-length of a segment at rest. On the other hand, the measurement of a segment with a measuring rod that *moves relative to it* requires the formulation of a new concept. This formulation is a matter of definition. We shall choose this definition in such a manner that it leads to a reasonable extension of the concept of length. This definition reads: *The length of a moving line-segment is the distance between simultaneous positions of its endpoints.*

The technique of extending a concept which we have used here can be illustrated by an elementary example. It corresponds to the introduction in vector analysis of the concept of the *geometrical sum* (vector sum) of two segments. The concept of the sum of two segments is first defined for segments lying in the same direction, i.e., as their *algebraic sum.* In Fig. 26 the sum of the segments AB and BC

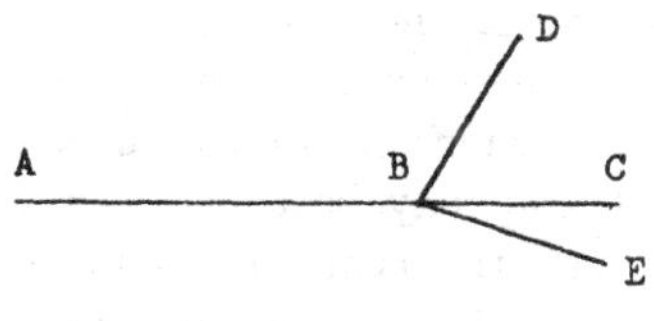

Fig. 26. The geometrical sum of two line-segments as an example of the extension of a concept.

is given by the distance AC, but we do not know how to compute the sum of the segments AB and BD, unless we introduce a new definition. It would be wrong to assume that this sum is given by the rotation of

BD into the position BC and by addition in the previous sense, i.e., by AC. This statement does not follow from the earlier definition, but represents a *new* definition which we shall call the *scalar sum*. We could just as well have given the new definition differently and specified the distance AD as the sum of the two segments AB and BD. This new sum is the *geometrical sum*. Is this new definition correct? The question is not reasonable, because the new definition does not follow from the old one and is therefore arbitrary. There is only one restriction which we shall impose, namely, that the new definition, which is more general than the old one, should coincide with it in those special cases where the old definition applies. This condition is satisfied by the geometrical sum, because it yields AC as the sum of AB and BC. The new definition is therefore a *consistent extension* of the old one. This condition, of course, is also satisfied by the scalar sum. The principle that such extensions must be consistent with existing definitions therefore does not lead to uniquely determined concepts, but leaves them arbitrary within certain limits. The choice of the geometrical sum instead of the scalar sum in vector analysis is based on other considerations. The geometrical sum yields different results for the additions $AB+BD$ and $AB+BE$, namely the distances AD and AE, whereas the scalar sum yields the same results in each case, namely AC. Hence the advantage of the first concept consists in its greater usefulness and has nothing to do with truth.

The same method of extension is used in the formulation of the concept of the length of a moving segment, where a concept is developed for a case not considered in classical kinematics. Again the rule of consistent extension is satisfied, since the length of the moving segment becomes identical with the rest-length in the case where it is at rest. The new concept therefore satisfies the only condition that can be imposed under these circumstances. Since this condition does not prescribe a unique extension, we cannot say that the new concept is true. It is useful, and may be considered a meaningful extension of the concept of length; this is all that can be required.

It can easily be shown that the length of the moving segment, as we have defined it, will depend upon the definition of simultaneity. Imagine that the segment is in a given state of motion relative to the coordinate system; then the length of the projection depends on the definition of simultaneity used in K. If the definition of simultaneity were changed, the length of the moving segment would also change. Assume that the projection of A' occurs at the same time as before, but the projection of B' a little later.[1] B' will have moved slightly to the right and B will be farther to the right; AB therefore becomes longer. Suppose now that simultaneity has been redefined to make the instants of the two projections simultaneous; then the new distance AB yields the length of the moving segment $A'B'$. This length is now greater than it was according to the first definition of simultaneity, and the measurements of space are therefore dependent on measurements of time.

We can also show that the length of a moving segment depends on its velocity. Assume that the simultaneity-projection is made of a rod moving with the velocity v relative to a system K in which simultaneity is defined according to Einstein's definition (1, § 19). Subsequently the rod is brought into the system K and placed next to the projection. Will it now coincide with the projection? To say that it does is not an *a priori* assertion, but a physical hypothesis that must be tested by experience. The theory of relativity denies this hypothesis; it maintains that the projection is shorter than the rest-length, and that this difference becomes greater with increasing values of v. This result cannot be justified now, but we can understand intuitively that it is possible; it expresses a property of rigid rods. We also recognize that the opposite assertion is likewise merely a hypothesis. In § 22 we formulated an absolute theory of time and showed that there would be an absolute simultaneity if signal velocities were unlimited. Even in this case it would constitute an additional hypothesis to assume that the projection of a moving rod, based on this simultaneity, equals its rest-length. *It follows from the nature of the extended concept of length that the length of a moving segment is generally different from its rest-length.*

Even this simple logical fact has led to considerations that raise the question of truth in the wrong place. It has been asked: which is the "true length" of the rod, its rest-length or its length when in motion? Evidently this question is unreasonable. Is the algebraic sum a "truer sum" than the vector sum? This is nonsense. The length of a moving rod is conceptually different from the length of a rod at rest and will therefore in most cases have a different measure. At most, one could ask whether there might be a greatest or smallest length of the rod. According to the Lorentz transformation, the rest-length is the greatest length. It is not the true length of the rod but merely its

[1] More precisely, the projection of A' is the same point-event as before, whereas the projection of B′ is a point-event which stands in the relation *later than* to the previously used corresponding point-event.

true rest-length. A different measure represents the true length of the moving rod. The concept of truth does not enter into the comparison of these two types of length. Let us consider, as an example of these logical relations, a rod AB viewed from a point P_1 (Fig. 27); it then

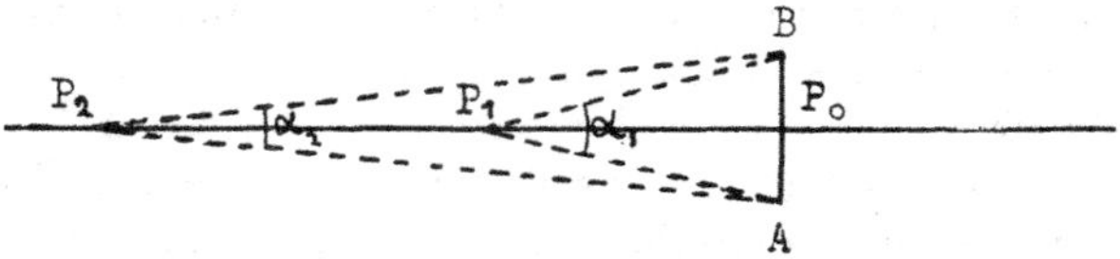

Fig. 27. The visual angle as an example of a relative magnitude.

subtends an angle α_1. The size of this angle changes when the point P_i travels along the horizontal, and has a smaller value α_2 at the point P_2. Which of these angles is the true visual angle? Again this question is unreasonable. A visual angle α_i is defined only for a given distance P_0P_i and there can be *a true visual angle only for such and such a distance.* There is also a largest subtended angle when P_i coincides with P_0 and $\alpha_i = \alpha_0 = 180°$. This however, is not the true, but the largest subtended angle; or, if we wish to characterize it by another property, it is the subtended angle for the distance zero. The extended concept of length that we have introduced has exactly the same logical structure. The length of a moving segment corresponds to the visual angle. It depends on the state of motion, just as the visual angle depends on the distance. The rest-length of the rod is its length at velocity zero, just as α_0 is the visual angle at distance zero. Both are maximum values.

There is an inessential difference between the two, viz., that the length of the moving segment depends not on one but on two parameters, on the state of motion of the segment and on the definition of simultaneity. If we call the length of the moving segment l, then l is a function of v and s, where v is the velocity and s symbolizes the definition of simultaneity. Thus we may write

$$l = l(v, s) \qquad (1)$$

This functional relation has the peculiarity that for $v = 0$ the function becomes independent of s, and

$$l(0, s) = l_0 = \text{constant} \qquad (2)$$

Indeed, if $v = 0$, the projection of B' (Fig. 25) upon K at a later time. is still B and the definition of simultaneity no longer has any influence

For this reason the rest-length of a body can be defined without reference to simultaneity. After we have introduced the extended concept of length which contains the concept of simultaneity, the special case of the rest-length must, strictly speaking, also contain the concept of simultaneity. The rest-length must therefore also be defined as the distance between the simultaneous positions of the endpoints. The resulting value, however, does not depend on the definition of simultaneity; every definition of simultaneity gives the same result for the rest-length and we can thus omit this addition in the definition of rest-length.

A rigorous analysis of these considerations leads to a correction of the rule of consistency for extended concepts used frequently in mathematics. One cannot say that, in the special case, the wider concept becomes *identical* with the narrower concept, but only that it *yields the same result*. It makes a difference whether we specialize a concept by giving one of the variables a special value, or whether the dependence on this variable is not even contained in the definition. Thus $l(v, s)_{v=0}$ is not identical with l_0, but leads to the same distance in the coordinate system. Similarly we must not say that the concept of the geometrical sum is identical with that of the algebraic sum for equidirected distances; it simply leads to the same numerical result.

Relation (1) formulates the famous assertion by Einstein [1] that the length of a rod depends on its velocity and on the chosen definition of simultaneity. There is nothing mysterious in this relation, for it is based on the fact that we do not measure the moving rod, but its projection on a system at rest. How the length of this projection depends on the definition of simultaneity can best be illustrated by reference to a photograph taken through a focal-plane shutter. Such a shutter, which is necessary for very short exposures, is not located between the lenses, but immediately in front of the film. It consists of a wide band with a horizontal slit, which slides down vertically. Different bands of the image are photographed successively on the film. Moving objects are therefore strangely distorted; the wheels of a rapidly moving car, for instance, appear to be slanted. The shape of the objects in the picture will evidently depend on the speed of the shutter. Similarly the length of the moving segment depends on the definition of simultaneity. One definition of simultaneity differs from another because events that are simultaneous for one definition

[1] Einstein's formulation usually is different, because he uses only the special definition of simultaneity (1, § 19) for each coordinate system, and consequently the state of motion of the segment determines at the same time the definition of simultaneity which is to be used. We therefore have $s = s(v)$, which reduces (l) to $l(v)$; thus the length of a moving segment depends only on its velocity.

occur successively for another one. What may be a simultaneity projection of a moving segment for one definition is a "focal-plane shutter photograph" for the other. A picture obtained by means of the first definition therefore appears distorted according to the second definition. The comparison with the focal-plane shutter makes this distinction clear, with the only difference that the focal-plane shutter operates much more slowly. It is a physical mechanism functioning at a speed below that of light, whereas the events of the projection form an unreal sequence with speeds above the velocity of light. (Cf. § 23). This difference is not important, however, for the qualitative description, and the changes in the length of moving rods can easily be visualized.

We emphasize this manner of visualizing the relation between space and time measurements because many presentations describe this relation in a misleading way. These presentations are based on the remark by Minkowski [1]: "From now on the ideas of space and time as independent concepts shall disappear and only a union of the two shall be retained as an independent concept." It is true that we may speak of a union of space and time in the relativistic theory of space and time. We shall discuss this idea in detail in § 29, p. 188. The first part of Minkowski's remark has unfortunately caused the erroneous impression that all visualizations of time as time and of space as space must disappear. The relativity of simultaneity does indeed lead to the coupling of space and time measurements and brings about a union of space and time, but this statement says no more than was expressed with the aid of the concept of the simultaneity projection and illustrated by reference to the focal-plane shutter. To make the union of space and time more apparent, we can use the following example. Let us consider a space filled with moving mass points, e.g., a gas the molecules of which are whirling about. At a given time each molecule will have a definite position. If we now change the definition of simultaneity, we shall obtain the same position for some of the molecules, but not for others; this distribution is now a "focal-plane shutter photograph." *The state of a space at a given time* is therefore not determined in an absolute sense, but depends on the definition of simultaneity. This situation can be visualized completely and represents everything that is asserted by the relativity of simultaneity about the union of space and time.

[1] Lecture given at the 80th *Versammlung Deutscher Naturforscher und Ärzte* in Cologne, Sept. 21, 1908.

The importance of the effect of the definition of simultaneity emerges even more clearly when we consider not only the length of a segment or the position of a point but also the *shape* of a moving object. Since the shape is determined by the simultaneous projection of all the points, it will evidently depend on the velocity of the object and on the definition of simultaneity. According to the Lorentz-Einstein assumptions, a moving circle assumes the shape of an ellipse, the minor axis of which lies in the direction of motion. This ellipse results from the fact that the Lorentz transformation calls for a contraction only in the dimension of the rod parallel to the line of motion. Such a deformation is not unintelligible; it merely shows that the simultaneity projection of the points of a moving circle has the shape of an ellipse in the coordinate system at rest. The idea of the focal-plane shutter photograph elucidates this assertion.

With this analysis all difficulties disappear that are intuitively anticipated with respect to the relativity of simultaneity and the distortion of moving bodies. One must only keep in mind that simultaneity is a matter of definition. Whether we are to associate the time order of distant events with the idea of the "click-click" of the focal-plane shutter or with the notion of the "tick" of a clock, is not determined by the events themselves. Both ideas apply directly only to experiences; the "click-click" is the experience of a sequence and the "tick" an instantaneous experience. We actually never experience distant events, but only the effects that reach us. Consequently we can choose how to coordinate these events to our visual images.

§ 26. CONSEQUENCES FOR A CENTRO-SYMMETRICAL PROCESS OF PROPAGATION

The dependence of spatial measurements on the definition of simultaneity has a peculiar consequence for processes of propagation such as light and sound waves, that travel from one center in all directions. It turns out that the shape of a single impulse depends on the definition of simultaneity in such a way that it is impossible to recognize the point from which the impulse has originated. For the sake of simplicity we shall speak in the following only of the motion of light. Similar arguments apply, however, to sound and any other centro-symmetrical process of propagation. In this analysis, the nature of the wave is not important—only the fact that it progresses in time.

Let us consider (Fig. 28) a single impulse of light produced at A at the time $t_1 = 0$. What shape does it have at time t_2? The answer depends on the definition of simultaneity. Let us first use the expression (1, § 19) and set ϵ equal to $\frac{1}{2}$ in

$$t_2 = t_1 + \epsilon(t_3 - t_1) \qquad 0 < \epsilon < 1 \tag{1}$$

We then put a clock at every space point, synchronizing all clocks from A according to this definition. Let us now imagine that all those points are marked that the light impulse has reached when their clocks show the time t_2; they are located on a circle around A. The solid circles in Fig. 28 thus indicate the position of the light impulse at

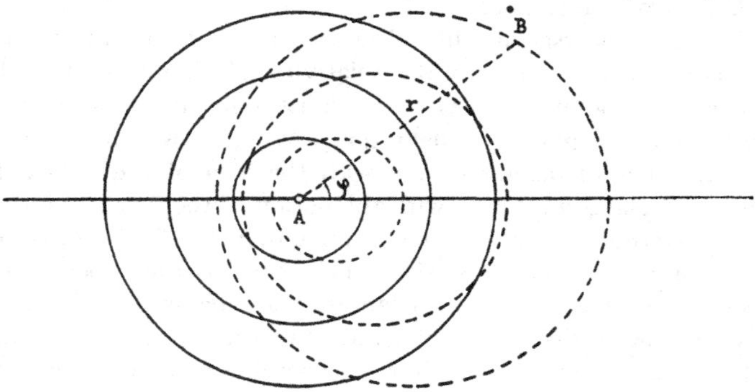

Fig. 28. Redefinition of a single light impulse from a centro-symmetric to an eccentric process of propagation.

different times t_2. The figure is drawn only for a plane, and the circles are to be understood as cross-sections of spheres. The light is propagated in spherical surfaces around A.

We shall now choose a different definition of simultaneity. We leave the clock at A unchanged, but avail ourselves of the choice which we have in setting the other clocks. We again use the expression (1) but let the factor ϵ be different from $\frac{1}{2}$ and make it dependent on the direction; we choose

$$\epsilon = \frac{c}{2(a \cos \phi + \sqrt{c^2 - a^2 \sin^2\phi})} \tag{2}$$

We set all our clocks from A according to this rule, and therefore the factor ϵ will be constant along every ray r. However, ϵ depends on the direction ϕ of the ray; the numerical value of ϵ is given by (2). In this formula a is an arbitrary constant and c is the numerical value of the velocity of light which was obtained by means of the first definition of simultaneity. The additional restriction on ϵ in (1) means only a limitation of the values of a, which can easily be fulfilled. Substituting (2) in (1), we have an admissible definition of simultaneity.

Let us call the newly defined time t'; the moment at which the light impulse departs from A is again $t'_1 = 0$. What is the location of the light impulse at time t'_2? If we let r be the length of the ray measured at the time t'_2 and if we then reflect the ray at the endpoint B of the ray r, it will return to A at a time t'_3; we then have $t'_3 = t'_3 - t'_1 = \frac{2r}{c}$. This measurement, which is made at A, does not require the t'-time and can therefore take the place of the expression obtained through the first definition of time. The arrival of the ray at B is now calculated from (1) and (2) as

$$t'_2 = t'_1 + \epsilon(t'_3 - t'_1) = \epsilon\, t'_3 = \epsilon \frac{2r}{c} = \frac{r}{a \cos\phi + \sqrt{c^2 - a^2 \sin^2\phi}} \tag{3}$$

The position of the light impulse at the time t'_2 is the curve given by this equation if we consider t'_2 as constant. With the substitutions

$$\begin{matrix} a\, t'_2 = e \\ c\, t'_2 = R \end{matrix} \qquad \cos\phi = \frac{x}{r} \qquad r^2 = x^2 + y^2 \tag{4}$$

of which the first two are abbreviations and the other two represent the change from polar to rectangular coordinates, (3) can now be written as:

$$(x-e)^2 + y^2 = R^2 \tag{5}$$

The location of the light impulse at the time t'_2 is again along a circle, but the center of this circle is displaced by the distance $e = at'_2$ along the horizontal from A. The family of circles which result for increasing values of t'_2 is indicated by the dotted circles in Fig. 28. The center of these circles is not stationary but moves to the right with the velocity a, while their respective radii R increase with the velocity c. For three-dimensional processes, these circles likewise should be regarded as the cross-sections of spheres.

The following results obtain. The surfaces of the propagation of a light impulse do not define a center. Depending on the definition of simultaneity, either the stationary point A, or a point that moves uniformly relative to A with the velocity a, can be regarded as the center of the light impulse. In either case the velocity of light equals c. This result is the basis of Einstein's principle of the constancy of the velocity of light; the motion of light can be considered as a spherical wave for any uniformly moving system.

A further explanation is required. What we have proved applies only to single impulses of light; only in this case is no center defined. However, a continuous source of light would define a center. If this constant source of light rested at A, the surfaces of propagation of the various light impulses would be the solid-line circles of Fig. 28, according to the first definition of simultaneity. The second definition, however, would give us not the family of dotted circles, but a system of such families of circles. These circles could then no longer be regarded as having one moving center, but each individual impulse would have its own center which is traveling to the right and is at A at the particular time when the impulse is sent. Each individual impulse would then have surfaces of propagation which, with increasing time, would be of the type of the dotted circles of Fig. 28. Consequently the state of motion of A could here be determined. Fig. 28 must therefore be conceived as representing a single impulse of light at different times, not as representing a periodic sequence of light impulses at the same time. This characterization of the state of motion of a constant source of light corresponds entirely to the conceptions of the theory of relativity. It is possible to identify by the type of propagation of light the system relative to which a constant source of light is at rest, for instance, by the Doppler effect which appears for another system and is quantitatively different at opposite sides of the source of light. The principle of the constancy of the velocity of light maintains only that the continuous motion of light has the form of a concentric spherical wave relative to that coordinate system in which the source of light is at rest. In the language of Minkowski: every cone of light $ds^2 = 0$ has a vertex; the line connecting the vertices of a number of such cones is the world-line of a point in a certain state of motion. The theory of relativity agrees with the classical theories in the statement that the constant source defines a center for all impulses. Relativity theory adds only that the propagations of all the impulses can be considered as spherical, measured from the system in which the source is at rest, provided that the definition of simultaneity is adjusted to the state of motion of the source of light.

It should be noted that the time transformation given in (2), which can also be expressed as

$$t' = t + \frac{r}{c} \cdot \frac{c - a\cos\phi - \sqrt{c^2 - a^2\sin^2\phi}}{a\cos\phi + \sqrt{c^2 - a^2\sin^2\phi}}$$

is not the Lorentz transformation. For the Lorentz transformation spatial measurements are also changed, because they are obtained relative to a moving system. In our example only the time was transformed, while the distances between points at rest remained the same; the spatial coordinates, therefore, retain their identity.

Our presentation enables us to visualize Einstein's result. The shape of the surface of the light wave is not uniquely determined but depends on the definition of simultaneity. Its so-called shape is always the simultaneity projection of the moving light impulse on a coordinate system. If we choose the first definition of simultaneity, we obtain the solid-line circles of Fig. 28. But if we imagine a focal-plane shutter photograph of these circles, where the focal-plane shutter moves from left to right, the dotted circles will result. This can easily be visualized, The farther the focal-plane shutter moves to the right, the later it will catch the light surface, and the dotted circles are therefore shifted to the right. Nevertheless it is not permissible to say that the second definition of simultaneity is *false.* The dotted circles appear like a focal-plane shutter photograph only relative to the first definition. Relative to the second definition all the projected events are simultaneous, and the solid-line circles form the picture of a focal-plane shutter photograph, produced by a focal-plane shutter which runs from right to left. This can also be visualized; for this purpose one must begin the analysis with the dotted circles.

§ 27. THE CONSTRUCTION OF THE SPACE-TIME METRIC

After these preliminary investigations concerning the connection between the definition of simultaneity and measurements of space. we now turn to the central problem of the physical theory of space and time, namely the complete construction of the space-time metric. We shall relate this construction to the causal theory of time which previously provided the definitions of time order and of time comparison (§ 21, § 22), and we shall show how metrical definitions can be added to these topological coordinative definitions in such a way that a physical space-time geometry results that is based entirely on the concept of the causal chain. We shall carefully distinguish between empirical statements and definitions and show to which statements of the relativistic theory of space and time these two categories apply. As before, the empirical statements are called "axioms" because they play the role of logical premises in the system of space-time theory.[1]

Let us imagine that numerous mass points are whirling about at random in empty space. On each of these points there is an observer and these observers can communicate with each other by signals.

[1] We shall present at this point only a summary. For a complete presentation we must refer to A. The numbers of the axioms and definitions in the following pages are given in accordance with those in A.

With the aid of these signals they now want to establish a space-time order.

For their measurements, the observers will use *first-signals*, i.e., light signals, because a time order constructed in this manner has the advantage that it cannot violate the relations *earlier* and *later* for other signals, since other signals are slower. The observers now have the task of choosing a system of points that can be called "at rest relative to each other" in order to define a space-time metric in this "rigid system." The choice of such a rigid system is again arbitrary and the state of relative rest is a matter of a coordinative definition. However, one definition is distinguished by its simplicity and leads to Newton's inertial system. That there exists such a definition is a matter of fact and can be either true or false. We must therefore state the basic empirical facts in the form of axioms. These axioms are satisfied, not under all conditions, but only in gravitation-free spaces. On the other hand, the applicability of these axioms provides a criterion for gravitation-free spaces. It is thus not necessary to emphasize that our presentation is restricted to gravitation-free spaces. This restriction is implied when the validity of the axioms is assumed.

The construction of the rigid system, or system of points at relative rest, is accomplished by steps which we shall describe in terms of certain operations performed by the observers on the various points. The observer at A knows what is meant by "temporal order at A," but he does not know as yet what is meant by "two equal successive time intervals." Provisionally, he lays down a completely arbitrary rule; i.e., he chooses a measure of time that differs by some monotonically increasing function [1] from the " uniform time" which is to be defined later.

He now tries to reach by signal a point B which has the property that the time interval $\overline{ABA}$ of a light signal [2] reflected at B will always have the same length when measured repeatedly at A in the arbitrarily chosen metric. $\overline{ABA}$ is therefore constant. To make this possible the point B must have a specific state of motion relative to A, which depends on the time metric chosen in A. Using this method, the observer at A looks for a number of such points $B, C, \ldots$ which form a "system related to A."

[1] A monotonically increasing function $y = f(x)$ is a function such that y always increases with increasing values of x, though the rate of increase may vary irregularly.

[2] The line indicates that the expression refers to the time interval and not to the spatial distance.

The observer at A can now transfer his time measure to the other points $B, C, \ldots$, for example, by giving a time signal every "second." The observers at $B, C, \ldots$ simply consider as equal those time intervals marked off by the arrival of the time signals from A. (This procedure does not yet require a definition of simultaneity.) These time intervals need not be equal to those at A in the ordinary sense, since nothing has been specified so far about the state of motion of $B, C, \ldots$ relative to A. We merely give provisional definitions.

The observers at $B, C, \ldots$ now make the following experiment. They measure whether the light signals BAB, BCB, CAC, etc., furnish also $\overline{BAB}$ = constant, $\overline{BCB}$ = constant, $\overline{CAC}$ = constant, etc., when they use for their time measurements the time scale previously supplied by A. Generally they will find that this is not the case. Even if they should not use the metric supplied by A, there would not be a metric in B which for all points yields $\overline{BAB}$ = constant, $\overline{BCB}$ = constant, etc. In other words: The "system related to A" is not a "system related to B."

Let us use this idea in order to combine a selection of points into a special system. For this purpose, we demand that a system S of points be chosen in such a way that this system is, *for each of its points, a related system.* That such a system exists is an empirical fact (Axiom IV, 1); that we select this system among all others represents an arbitrary definition.

Using the points of the system S obtained in this manner we now perform the following experiment. We send one light signal along the triangular path $ABCA$, another in the opposite direction along $ACBA$, and test whether the time interval $\overline{ABCA}$ equals the time interval $\overline{ACBA}$. Again, we do not yet employ the concept of simultaneity for distant points. We send the two signals simultaneously from A and observe whether or not they return to A at the same time. Generally this condition will not be satisfied for an arbitrary system S. Therefore we add a further restriction if we demand that a system S' is to be chosen among the systems S which satisfies the round-trip axiom. That such a system S' exists is again a matter of experience. (Axiom IV, 2).

So far we have not made use of the time comparison between the moving mass points, but it would have been possible to do so by means of the definition (2, § 19)

$$t_2 = t_1 + \epsilon(t_3 - t_1) \qquad 0 < \epsilon < 1 \tag{1}$$

since this definition contains no presupposition regarding the relative state of motion of these points. We have only to imagine that the clock at B is continually set relative to the clock at A, for we cannot construct a mechanism at B which, due to the arbitrariness of the time metric at A and the arbitrariness of the relative state of motion of A and B, will permanently maintain the once-established simultaneity. This conception does not involve any difficulties, and the resulting simultaneity will always satisfy the basic topological requirement that it connect only events which are indeterminate as to time order.[1]

Having made a choice among the mass points and having combined a system S' of these points into a spatial coordinate system, we can now ask whether a special simultaneity can be defined for this system. This simultaneity definition is given by setting ϵ equal to $\frac{1}{2}$. Its advantages consist in the following properties:

1. If a clock at B is set from A according to (1) and $\epsilon = \frac{1}{2}$, and if we then set the clock at A from B, using the *same rule*, the two times agree. (The synchronization is symmetric.)
2. If two clocks at B and at C are set from A according to (1) and $\epsilon = \frac{1}{2}$, and if the two clocks are compared directly with each other by the *same rule*, they will be found to agree. (The synchronization is transitive.)

These properties are by no means self-evident; they require the assumption of the previously mentioned light-axioms, which apply in S'. These properties make the time order in S' particularly simple and justify the definition of simultaneity used by Einstein in the special theory of relativity where $\epsilon = \frac{1}{2}$. This should not mislead us into believing that this definition is "more true" because of its simplicity. Again we are concerned with nothing but descriptive simplicity. The choice of a more complicated definition of simultaneity does not present any difficulties for our imagination. One should not confuse here "transitivity of simultaneity" with "transitivity of simultaneity according to the *same rule of synchronization.*" The first applies always, once a simultaneity has been uniquely defined for *one* clock according to (1). For the comparison of any given clocks a special value of ϵ, which may vary with time, will have to be chosen. The second kind of transitivity depends on certain physical conditions. Even if these are satisfied, we are not compelled to carry through this simple kind of simultaneity.

[1] This assertion requires a special proof; see A., § 7.

In addition to the simplicity of their time order, the systems S' possess a second important property: they permit us to make *spatial measurements.* This fact is of extraordinary significance because it proves that space measurements are reducible to time measurements. *Time is therefore logically prior to space.*

First we shall define an important topological concept of spatial order, namely the concept *between.* We treated the same concept in § 14 and showed there that the logical significance of this concept can be determined by an implicit definition. Here we have to achieve more: we have to discover physical relationships that can be coordinated to the concept *between* and thereby permit an *application* of this logical concept to physical *reality.* As we have explained in §§ 14 - 15, the logical concept does not prescribe any particular application. We shall here consider that application which leads to physical geometry. Therefore we must now give the *coordinative definition of the concept between.* Surprisingly enough this definition can be given solely in terms of temporal concepts. The topological neighborhood relations of space are therefore reduced to temporal relations and thus to causal relations.

Definition e. A point B lies *between* A and C, if the first-signal ABC arrives at C at the same time as the first-signal AC. (In short, if $\overline{ABC} = \overline{AC}$.)

If this definition is not to contradict the purely mathematico-logical significance of the concept between, the following empirical rule must be added:

Axiom G. If for two points B_1 and B_2 it is true that $\overline{AB_1C} = \overline{AC}$, as well as that $\overline{AB_2C} = \overline{AC}$, then either $\overline{AB_2B_1} = \overline{AB_1}$ or $\overline{B_1B_2C} = \overline{B_1C}$.

Having determined the concept between, we can now take a further step toward spatial measurements and define the concept *straight line,* which is a metrical concept. In this case we shall not employ the method of implicit definition used in § 14, but another logically permissible method, deriving the concept straight line from the concept *between* and the basic logical concept of a *set.*

Definition f. The straight line through A and C is the set of those points which among themselves satisfy the relation *between* and which include the two points A and C.

It must be shown that the straight line determined by this coordinative definition agrees with the geometrical concept of straight line, e.g., that the straight line from A to B is identical with the straight line from B to A, etc. This proof is possible on the basis of the axioms given thus far.

We shall now define the metric in its physical sense by giving the coordinative definition of spatial congruence:

Definition 10. If the time interval $\overline{ABA} = \overline{ACA}$, then the spatial distance AB is equal to the spatial distance AC. (See Fig. 29.)

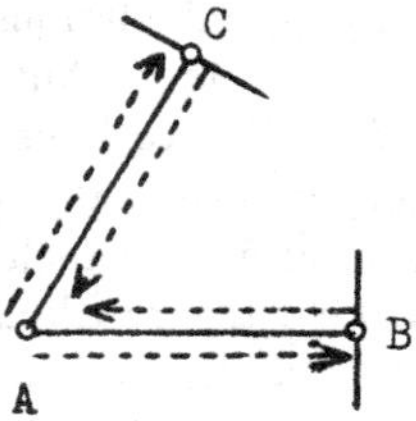

Fig. 29. Definition of the equality of spatial distances in terms of measurements of time intervals.

These are all the definitions required for a geometry of space. We can answer any question of a geometrical nature by using time measurements exclusively. We even gave overdeterminations and must now prove the consistency of our system, which is easily done. It can readily be shown that the straight line of definition *f* is also the shortest line in the sense of the metric of definition 10.

Let us imagine ourselves in one of the systems S' in which we intend to carry out measurements. Specifically, we are interested in determining the geometry of S'. Once a definition of congruence is given, the choice of the geometry is no longer in our hands; rather, the geometry is now an empirical fact.

If we were to measure the circumference and diameter of a circle, would their ratio equal π? In general the geometry will not be Euclidean. But we can select certain systems S'' from the systems S', if we demand that the selected systems satisfy the condition that the defined geometry be Euclidean. It is again an empirical fact that there are such systems S'' (Axiom V), whereas their selection is based on a definition.

Now we have reached an important goal: we have defined in each system of the class S'' a complete and unique metric without the use of rigid bodies or natural clocks. Light signals alone provide the metrical structure of the four-dimensional space-time continuum. This construction may be called a *light-geometry*. Its applicability depends on the truth of the previously mentioned axioms which refer only to light signals and mass points and are therefore called *light-axioms*.

On the other hand, a second goal has not yet been attained. We have not yet sufficiently restricted the class S'' of systems to make it identical with the class of Newtonian inertial systems I. The class S'' is still too general and contains other systems beside I. A further explanation is required. We started with freely moving individual mass points and constructed from them a point system K by placing certain restrictions upon the results of measurements performed by light signals. It can now be shown that there is not *one*, but there are many systems that satisfy these requirements; they form the class S'' It is not our purpose to define a class containing exactly one system, since the Newtonian inertial systems form a class I of systems that move uniformly relative to each other. It is the goal of our construction to arrive at this particular class I, i.e., to determine with the aid of light signals not only the geometry *within* a given system of points, but to establish at the same time a choice among the many point systems, such that only systems in a particular *state of motion* satisfy our requirements. We thus want to specify the state of motion in space, taking the motion of light as the physical "framework" to which the systems are "tied." The motion of light is too loose a framework, however, to determine the state of motion sufficiently to make it identical with that of the inertial systems. We obtain a more general class S'' which contains the inertial systems as a proper subclass. Yet a considerable restriction has already been imposed on the totality of possible states of motion, and a method has been indicated by which a state of motion can be defined in free space, provided that there exists a physical process, independent of the procedure of measurement, which can be utilized.

What constitutes the excessive generality of the class S''? This class may be understood mathematically as follows. If any system K of the class S'' is given, it is possible to transform it into another system K' belonging to the same class by means of a coordinate transformation

$$x_i = f_i(x'_1 \ldots x'_4) \quad i=1\ldots4$$

The class S'' will now have to be characterized by a specification of the exact form which this transformation will have to assume. The following considerations will achieve this aim.

The light-geometry has been constructed for each of the systems S'' and the equality of distances is so defined that a light ray travels equal distances in equal times, i.e., the velocity of light becomes constant. Since, according to our assumption, the light-geometry is Euclidean in each of these systems, the propagation of light is given by the relation

$$\Delta x_1^2 + \Delta x_2^2 + \Delta x_3^2 = c^2 \Delta t^2 \tag{1}$$

where Δ indicates the difference between corresponding coordinates of the two endpoints. If we write x_4 for ct (this is only a convenient notation for the unit of time), and put x_4^2 on the left-hand side, the equation reads

$$\Delta x_1^2 + \Delta x_2^2 + \Delta x_3^2 - \Delta x_4^2 = 0 \tag{2a}$$

In the moving system K', the propagation of light must be given by a corresponding equation of the same form

$$\Delta x'^2_1 + \Delta x'^2_2 + \Delta x'^2_3 - \Delta x'^2_4 = 0 \tag{2b}$$

The required transformation is therefore characterized by the condition that it transforms (2*a*) into (2*b*).

The solution to this problem is well known to the mathematician. The condition is satisfied by the linear transformations [1]

$$x_i = a^i_k x'_k \tag{3a}$$

where the coefficients satisfy the condition

$$[a^m_i a^m_l] = \begin{cases} +k^2 \text{ for } i = l = 1, 2, 3 \\ \quad 0 \text{ for } i \neq l \\ -k^2 \text{ for } i = l = 4 \end{cases} \tag{3b}$$

and the summation is made over the repeated superscript. These transformations are identical with the Lorentz transformation except for the constant k, which will be discussed presently. Another transformation that satisfies the condition is given by the form

$$x_i = \frac{x'_i}{x'^2_1 + x'^2_2 + x'^2_3 - x'^2_4} \tag{4}$$

These relations are called similarity transformations. If one coordinate system satisfying (2*a*) is given, the class S'' is then determined as the

[1] We are using here the customary notation, which omits the summation sign. Instead we are using the rule that one has to sum over every index that appears twice. Formula (3*a*) therefore represents four equations, the right side of each of which is a sum of four terms. In (3*b*) we use an extension of this notation. The summation is made over m, but the square bracket indicates the rule that when $m = 4$ the corresponding term is to be written with a minus sign.

totality of those systems derivable from the given one by the transformations (3) and (4).

Class S'' is therefore too general, since the class I of inertial systems is connected by the Lorentz transformations alone, while transformation (4) leads out of this class. The class S'' constitutes the limit for the determination of the state of motion by light-geometry. The determination of class I, a subclass of S'', is not possible unless we can avail ourselves of some physical means other than light signals. Only one argument can be raised in favor of defining class I by means of light-geometry alone; it can be shown that a system T which belongs to S'' but not to I has a physical singularity. The system T contains light signals that pass through infinity and yet return within a finite interval of time; and it contains finitely located points that can be connected by light signals in one direction *only* while the return signals never reach their goals. If we consider the difference between finite and infinite as physically recognizable in the sense of § 12, the class I can therefore be defined by means of the light-geometry. However, since the light-axioms apply, as we shall discuss later, only in limited regions of space, and since no unlimited spaces can be utilized for a decision, this method is not fruitful. We can always describe systems T that deviate from systems of class I only outside the space we have at our disposal.

We must therefore look for another way to exclude transformation (4) and thus systems of the kind T. If we go from a system K by means of the transformation (4) to a system K', the points of the latter are not at rest relative to the points of K, and its space axes constantly expand relative to K. Therefore K' has a different measure of time.[1]

We can therefore exclude these transformations by introducing material bodies. The points of a system I can always be connected by rigid rods, while this is not possible in a system T. Thus we have obtained a definition for inertial systems. Furthermore, the time of a system I corresponds to the time of natural clocks, which is not true for a system T. This reference to clocks could also be used as a definition of the class of inertial systems. With the introduction of material bodies it is thus possible to eliminate the systems T. It should be noticed that this method does not even utilize the most important function of material bodies. The rigid rod is not used for the definition of *spatial congruence* within the system, but only for the

[1] For a precise calculation of this case we must refer to § 16 in A. For a correction to A., see also *Zs. f. Phys.* 34, 1925, page 34.

determination of one—and indeed only one—distance as rigid, i.e., for the definition of the *temporal congruence of spatial distances.* For this reason natural clocks can be employed for the definition of rigidity. It suffices to specify the *time metric* at a space point. Then the rigidity of the entire system and of the set of systems, even their state of motion, is determined by means of light signals. Although this determination refers only to the *internal* state of the systems, it thus establishes their *state of motion in space.* This is only possible because the light-geometry introduces rather strong restrictions and connects every possible state of motion with a corresponding internal metric.

Finally we shall investigate how to change the transformations (3) into the Lorentz transformations, with which they are not yet identical because of the general constant k. We required of the desired transformations that they leave the expression (2) invariant; this requirement corresponds to the condition that the light-geometry be carried through in each system, but it does not yet determine the comparison of units between systems in different states of motion. Yet to be specified is the *comparison of the rest-length of moving line-segments,* the first comparison of length in kinematics (compare § 25). If rigid rods are used as the unit in K, then it is convenient to call the rest-length of the rod transported into the moving system K' equally long in the new system. This procedure is not possible in the light-geometry, however, since there is nothing to transport. Consequently another comparison of the units must be found; it is accomplished with the aid of the second comparison of length in kinematics. If we consider arbitrary units of length in K and K' respectively, then the unit in K can be measured in K', and vice versa, with the aid of the concept of *length of a moving segment.* We shall not obtain the same contraction or expansion factor in both of these cases because of the arbitrary choice of the units. If the identity of this factor is required by definition, however, the choice of the rest-unit of one will depend on the choice of the other, and we have thus given a rule for the first comparison of length in kinematics.

These conditions are identical with setting $k = 1$, as can easily be shown. Since coordinative definitions are arbitrary, we can set $k = 1$. With this additional rule, (3) becomes identical with the Lorentz transformation, which can now be written in the form

$$(5a)\quad x_i = a_k^i x'_k \qquad [a_i^m a_l^m] = \begin{cases} +1 \text{ for } i = l = 1, 2, 3 \\ 0 \text{ for } i \neq l \\ -1 \text{ for } i = l = 4 \end{cases} \quad (5b)$$

To go from here to the well-known form of this transformation, we need only a few simplifying specializations which again are of the nature of definitions. We give identical origins to the two systems, choose their space axes parallel, and specify the direction of motion of one system as being the direction of the x_1 axis of the other. This results in the familiar form of the Lorentz transformation:

$$x_1 = \frac{x'_1 + vt'}{\sqrt{1-\frac{v^2}{c^2}}} \quad x_2 = x'_2 \quad x_3 = x'_3 \quad t = \frac{t' + \frac{v}{c^2}x'_1}{\sqrt{1-\frac{v^2}{c^2}}} \tag{6}$$

in which "ct" and "ct'" are substituted for "x_4" and "x'_4" respectively. The twenty constants of transformation (5) are reduced by these special conditions to the single constant v.

An exact presentation of these special conditions is given in § 17 of A., where they are expressed by means of restricting equations on the a^i_k. Simple calculations lead from them to the special values of the a^i_k which appear in (6). The derivation of equations (6) from the invariance (2) can be found in many texts on the theory of relativity, but the distinction between definitions and empirical statements is usually not stated clearly.

The goal of determining the class of inertial systems and their metric has thus been attained, and this chapter of the theory of space and time might be regarded as closed. However, since there are also measuring instruments other than light, namely measuring rods and clocks, and since these have so far played a subordinate role in our construction, it is important to ask how these objects behave in relation to the light-geometry. We could have started with these measuring instruments and used light solely for the definition of simultaneity, in which case we would also have obtained a geometry. We may now ask how this geometry would be related to the light-geometry. The statements about this relation are formulated as the *matter-axioms* in contrast to the *light-axioms* used exclusively so far.

The formulation of these statements requires an additional consideration. Because of the definitions involved, the constructed light-geometry is arbitrary. The choice of the definitions determines whether the relativistic or the classical light-geometry will be obtained. The light-axioms can be the same for both. The relativistic light-geometry that we have developed above differs therefore from the classical theory only in the choice of definitions. Instead of the Lorentz transformation, Galileo's transformation could thus be defined. For this purpose, a spatial congruence that varies from system to system would have to be defined. Distances traveled by light in equal times would then in general not be equal, since according to the Galilean transformation the velocity of light depends on the direction of the moving system. Furthermore, another comparison of the rest-length of moving segments and another simultaneity would have to be defined, for which the factor ϵ in formula (1) depends on the direction. The *classical light-geometry* is possible, because the light-axioms in the theory of relativity do not differ from those in classical theory except for the assertion of the limiting nature of the velocity of light. Even if this axiom applies, however, the Galilean transformation can be defined. The difference consists solely in the fact that only systems moving slower than light can be realized by material things.

The context of the matter-axioms (axioms VI—X) can be summarized as follows: *Material things follow the relativistic light-geometry.* If distances which are light-geometrically equal are measured by rigid rods, they will also turn out to be equal. If the flow of time, light-geometrically defined as uniform, is compared with that of a natural clock, they are found to agree. Similar results hold for assertions concerning the transport of clocks and measuring rods into moving systems. We find agreement when they are compared with units which were light-geometrically transferred to the moving system. For this transfer we must use the definitions of the relativistic light-geometry, not the classical definitions. *Einstein's assertion can be expressed by saying that material things adjust, not to the classical, but to the relativistic light-geometry.*

This assertion constitutes what is new in the theory of relativity from the point of view of physics. Whereas all of the light-axioms hold in classical optics, to which the theory of relativity adds only the assertion that the velocity of light is the upper limit for the speed of signals, the matter-axioms signify a deviation from classical theory. They contain the assertion that the Lorentz transformation, which in the light-geometry differs from the Galilean transformation only *by definition,* is at once the transformation for measuring rods and clocks. This assertion contains, therefore, that part of the relativistic theory of space and time which is to be tested empirically.

We have now succeeded in distinguishing between the physical assertions of the relativistic theory of space and time and its epistemological foundation. This epistemological foundation is supplied by the discovery that coordinative definitions are needed far more frequently than was believed in the classical theory of space and time, especially for the comparison of lengths at different locations and in systems in different states of motion, and for simultaneity. The physical core of the theory, however, consists of the hypothesis that natural measuring instruments follow coordinative definitions different from those assumed in the classical theory. This statement is, of course, empirical. On its truth depends only the *physical theory of relativity*. However, *the philosophical theory of relativity*, i.e., the discovery of the definitional character of the metric in all its details, holds independently of experience. Although it was developed in connection with physical experiments, it constitutes a philosophical result not subject to the criticism of the individual sciences.

In the following we shall demonstrate how the content of the light- and matter-axioms can be visualized geometrically by the world-geometry of Minkowski.

§ 28. THE INDEFINITE SPACE-TYPE

We have derived the Lorentz transformation by transforming the expression

$$\Delta x_1^2+\Delta x_2^2+\Delta x_3^2-\Delta x_4^2 = 0 \qquad (1)$$

into a similar expression in the variables x'_1, a transformation which leaves expression (1) *invariant*. It can be shown that the Lorentz transformation (5, § 27) possesses the additional property that it also leaves the expression

$$\Delta x_1^2+\Delta x_2^2+\Delta x_3^2-\Delta x_4^2 = \Delta s^2 \qquad (2)$$

invariant. This is a special property of the linear transformation (5, § 27) not shared by transformation (4, § 27).[1] The Lorentz transformation can therefore be exhaustively defined in a purely mathematical fashion, if we specify that it leave expression (2) invariant, which condition automatically determines the value $k = 1$. This property is the basis of Minkowski's geometrical interpretation of the Lorentz transformation.

This interpretation is carried through on the basis of the formal analogy of (2) with the Pythagorean theorem. We shall concern

[1] The latter transforms the left-hand side of (2) into the same expression in the variables x_1, which however is multiplied by a factor $\lambda(x'_1 \ldots x'_4)$. Only if the right-hand side equals 0, as in (1), this transformation leaves (2) invariant, since we can divide by λ in this case.

ourselves first with the extension of the space-type which results from these considerations; its application will follow in § 29. This extension starts with the concept of length.

By the length of a line-segment we do not understand the segment itself, but a number coordinated to it. A segment is determined if we give the coordinates of the totality of points which lie on it. This information does not tell us, however, how long it is; an indication of length signifies in addition a comparison of this segment with other segments. Length is therefore determined only if a definition of congruence is added. The definition of congruence is given in analytical geometry by a formula correlated to the coordinates of the segment, which determines the length of the segment as a function of the coordinates. For a straight segment it would be sufficient to know the two endpoints, in which case formula

$$\Delta x_1^2 + \Delta x_2^2 + \Delta x_3^2 = \Delta s^2 \qquad (3a)$$

which is the same as the Pythagorean theorem, expresses the distance for three-dimensional space. Incidentally, this simple form can be used only if the coordinates satisfy certain conditions; they must be rectilinear and orthogonal. A generalization to arbitrary coordinate systems will be discussed in § 39. Extending the theorem to a four-dimensional space, we would have

$$\Delta x_1^2 + \Delta x_2^2 + \Delta x_3^2 + \Delta x_4^2 = \Delta s^2 \qquad (3b)$$

This formula, however, is not yet equivalent to expression (2), in which the term Δx_4^2 has a negative sign. How can our considerations be extended to such an expression?

The extension is possible if we employ the previously discussed rule of consistent extension. We shall define the concept of length in such fashion that it will include expression (2), as follows:

Definition. The coordination of a number Δs to the coordinate differences $\Delta x_1 \ldots \Delta x_4$ by means of the *fundamental metrical form*

$$\Delta x_1^2 \pm \Delta x_2^2 \pm \Delta_3^2 \pm \Delta x_4^2 = \Delta s^2 \qquad (4)$$

defines the measure of length.

This extension of the concept of length represents a logical procedure previously (§ 25) explained by the example of the concept of the vector sum. It cannot be called true or false since it merely constitutes a definition. It satisfies the rule of consistent extension because it becomes identical with (3) in the special case when all the signs are the same. A metric of this special kind is called *definite,* while one that has positive as well as negative signs is called *indefinite.*

The difference between a definite metric and an indefinite metric of the form (2) can best be understood by means of Fig. 30*a* and *b*. For the sake of simplicity these figures are drawn for only two coordinates, and we have

$$(5a) \qquad \Delta s^2 = \Delta x_1{}^2 + \Delta x_4{}^2 \qquad\qquad \Delta s^2 = \Delta x_1{}^2 - \Delta x_4{}^2 \qquad (5b)$$

Figure 30*a* corresponds to the definite type (5*a*). The lines which are at a constant distance from the origin, the contour lines, are circles,

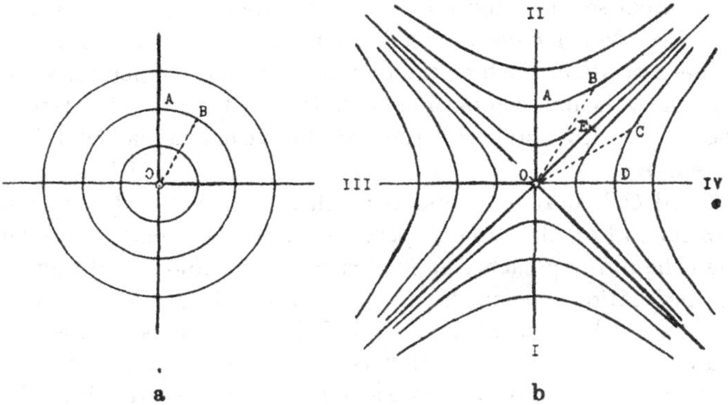

Fig. 30. Definite and an indefinite metric.

since (5*a*) represents the equation of a circle with the radius Δs. Fig. 30*b* corresponds to the indefinite type (5*b*); the contour lines are now hyperbolas, since (5*b*) is the equation of a hyperbola, the distance of whose vertex from the origin is Δs. The hyperbola $\Delta s^2 = 0$ has degenerated into the two asymptotes.

Now (5) defines a metric and Δs therefore is to be conceived as the distance of the point (x_1, x_4) from the origin. This statement is immediately clear for Fig. 30*a*, which simply states that *OA* equals *OB*. The same must be true for Fig. 30*b*, and consequently every point on a given hyperbola must have the same constant distance from the origin; *OA* must equal *OB* also in Fig. 30*b*. While this equality corresponds to the normal congruence for Fig. 30*a*, we must remember the definitional character of congruence in order to understand Fig. 30*b*. The distances *OA* and *OB* in Fig. 30*b* appear different because (§ 11) our visual estimate is adjusted to the behavior of rigid bodies, which satisfy the relations of Fig. 30*a*. We can now adopt the previously used method and visualize the new metric by imagining bodies that would satisfy it. Let us therefore imagine the distance *OA* realized by a rod. If we now rotate this rod so that one end remains fixed at *O*, the other end will travel along the hyperbola and the rod will assume the position *OB* and similar positions. We can easily visualize this situation if we remember that congruence is a matter of definition. The behavior of the imagined rods thus offers a visual realization of the metric (5*b*).

The difficulties arising from the metric of Fig. 30*b* do not spring from the comparison of length but from another property of this metric. The lines which have an inclination of 45° are *null lines*, i.e., any distance on them, such as the distance *OE*, has the length zero. This singularity is not compatible with the customary visual concept of length, yet this property too can be visualized as will be shown presently.

A rod *OB* rotated around *O* reaches the position *OA* and remains with its end on the same hyperbola, so that the end moves far away when the rod approaches an inclination of 45° and lies at infinity when that inclination is reached. In this position the rod would be infinitely extended. We shall assume, as in § 12, that this is impossible, which means that the 45° inclination is a limiting position that cannot be reached but can only be approached as closely as we wish. We shall assign to a segment *OE* in this singular position the length zero, because the asymptote represents the innermost (degenerate) hyperbola and corresponds therefore to the contour line of the shortest measure of length.

On either side of the limiting position, the rods behave normally according to the hyperbolic contour lines. Because of the unattainability of the limiting position, however, it is not possible to rotate a rod from the position *OC* into the position *OA*. Instead, it can be brought into the position *OD*. We have therefore two kinds of rods, which we shall call the blue and the red rods. The direction of the blue rods lies always in quadrants I and II, that of the red ones always in quadrants III and IV. It is possible to move the rods so that they intersect the asymptotes, but they can never be rotated beyond the limiting position. A red rod can therefore never be brought into the position of a blue rod and vice versa. To distinguish between these two kinds of length, we shall give the Δs^2 of quadrants I and II a negative sign.

The peculiar form of metric that admits of lengths $\Delta s^2 = 0$ and $\Delta s^2 < 0$ is the natural expression of the behavior of the described measuring instruments. The division into two types of rods which cannot be compared is symbolized by the distinction between $\Delta s^2 > 0$ and $\Delta s^2 < 0$, and the limiting case by $\Delta s^2 = 0$. This metric entails certain deviations from the customary concept of length; but these deviations are necessitated by the behavior of the described rods. We shall now follow this line of thought in more detail.

It was mentioned before that we understand by the length of a segment, not the segment itself, but a number coordinated to it. The customary measurement of length, although it recognizes this distinction, adds the further requirement that the length zero be coordinated only to a point. This requirement is violated for the limiting position (but only for this position). The concepts of *length* and *extension* no longer coincide in this special case. A segment in the limiting position can be *long* in the sense of *extended*, and yet have the length zero. Extension is a topological concept; a geometrical structure is *extended* if it contains a continuous sequence of points. Whether or not there is an extension is determined by the coordinate system; extension is therefore a concept referring to coordinates. What *length* is to be coordinated to a geometrical element, however, is not determined by the coordinate system. Instead, the measurement of length characterizes the behavior of measuring instruments. This behavior includes a singularity in the limiting position; a rod of finite length will not coincide with a finite line-segment in the limiting position, but compared to it would be infinitely long. Because of this peculiarity we assign to every finite extension in the limiting position the length zero. If we were able to rotate an extension like OE into a different direction, it would assume the extension zero. In renouncing, for the limiting position, the requirement that length zero is to be reserved exclusively for a point, we adjust our metric to the behavior of our measuring rods, relative to which the measure of a finite extension in the limiting position is indeed zero. We have here a topologically different behavior of measuring instruments and are therefore forced to abandon the mentioned requirement although it is satisfied by the customary measurement of length.

It is merely a reasonable extension of the concept of length to admit negative numbers for the square of a length. Customarily, length is defined as a positive number, but there are instances where the introduction of negative length appears useful. In the metric of Fig. 30*b* even the square of a length can become negative, and therefore the length itself imaginary. But this fact is of secondary importance, because it is irrelevant whether we consider Δs or Δs^2 as the length. The distinction is merely a matter of expediency. As always, the introduction of imaginary numbers is merely an arithmetical device, which by no means implies that "space or the measuring instrument becomes imaginary." That would be nonsensical.

Our presentation has shown that a manifold of the indefinite type can be visualized just as well as one of the definite type. If it is said that a nonvanishing extension of the length zero, or a length whose square is negative, cannot be visualized, such an objection indicates that requirements have been tacitly retained which apply only to the usual measuring instruments. It is not the advantage of visualizability that makes us retain these requirements for the customary geometry of space; it is merely the behavior of the usual measuring instruments that induces us to project a metric of the definite type into the space of our environment. If the measuring instruments actually behaved like the blue and red rods, their behavior would be represented by the indefinite type of metric in the same sense as we can say that the behavior of the usual measuring rods is characterized by the definite type of metric. We have thus visualized the indefinite type of space by imagining the corresponding behavior of measuring instruments.

At the same time, we recognize that the customary measuring instruments represent a special type of geometrical relations. These results may be added to our previous considerations regarding the problem of space, although the ideas expressed in this section are intended to serve mainly for the application of the indefinite metric to a quite different manifold, namely, the space-time manifold.

§ 29. THE FOUR-DIMENSIONAL REPRESENTATION OF THE SPACE-TIME GEOMETRY

We have illustrated in the previous section the indefinite space-type by means of the behavior of rigid rods, considering them as the realization of the Δs^2. Thus we described it as a pure *space*-type. In the case of the Lorentz transformation, however, the fourth dimension is given by *time*, and the realization of Δs^2 must therefore be carried out in a different fashion. We shall discuss this method in connection with the work of Minkowski.

Let us recall a consideration of § 16 according to which the time dimension differs basically from the space dimension. If we wish to study the actual space-time manifold, we must therefore look for *space-time objects* that behave like the distances OA, OB . . ., etc. Such objects are clocks and, with an additional qualification, *measuring rods*.

The point of the space-time manifold is the *point-event*, i.e., an event determined by three space coordinates and one time coordinate. The "tick" of a clock is an example of such an event. A space point P, on the other hand, is represented by a line which corresponds to the flow of time at P and is called its world-line. The space point at rest is represented by a vertical line; a slanted straight line would correspond to a uniformly moving space point, since its position in the coordinate system changes with time. Any two point-events determine a "distance" Δs. We shall call such a distance an *interval*, in contrast to spatial distances and, as we shall see, also in contrast to temporal distances. The interval is therefore a metrical concept which corresponds to length and not to extension and is determined by its two endpoints.

Let us look first for the interval $\Delta s^2 = 0$. This interval is given by the motion of light, which satisfies the equation $\Delta s^2 = 0$. According to page 179 this equation is represented by the asymptotes of Fig. 30*b*, and the world-line of a light ray traveling along the axis is therefore given by a straight line having an inclination of 45°. For this light ray $x_1 = x_4$, since we have set the velocity of light equal to 1.

The limiting position in Fig. 30*b* is thus represented by the motion of light. World-lines whose directions lie in quadrants I and II are called *timelike* (Fig. 31); they correspond to space points that move *slower* than light. The slope of the straight line is a direct measure of the velocity of the corresponding space point; the closer its inclination approaches the asymptote, the greater will be its velocity. World-lines whose directions fall into quadrants III and IV are called *spacelike*. They cannot be realized by moving mass points, since they would require a velocity greater than that of light. The difference between the blue and red rods in our illustration of § 28 corresponds therefore to the difference between timelike and spacelike world-lines. Timelike world-lines, according to our definition of time order in § 21, are those lines whose point-events *follow one another in time*, since they can be connected by means of signals. Spacelike world-lines, however, connect point-events that are *indeterminate as to time order* (§ 22).

The choice of the coordinate axes has the following significance. We may choose any timelike line as the time axis, because this is compatible with the definition of time order. Since differently inclined timelike lines represent the world-lines of mass points in different states of

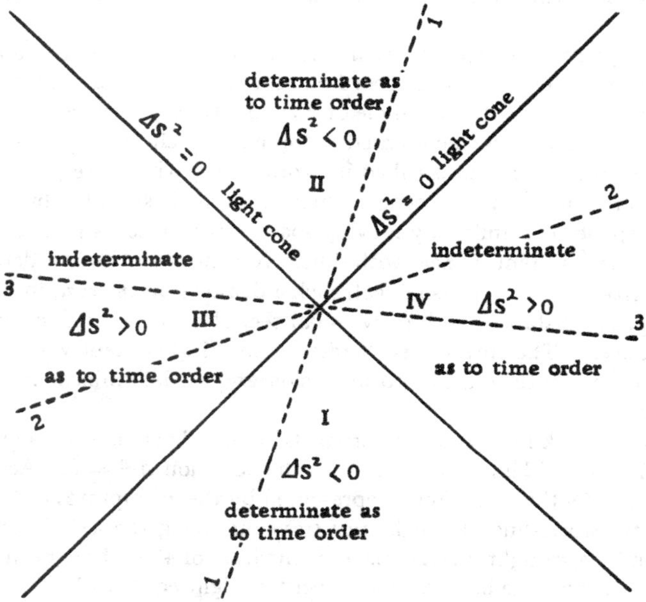

Fig. 31. Division of the space-time manifold.

motion, the choice of one of them as the time axis constitutes the selection of one state of motion as the state of rest. The choice of the time axis therefore determines the state of motion of the coordinate system. Any spacelike line, on the other hand, may be chosen as the space axis, since we can always define its point-events as simultaneous. The choice of a definite spacelike line as the space axis represents therefore the choice of a particular definition of simultaneity.

Generally speaking, any space axis may be combined with any time axis to form a coordinate system, as in Fig. 31 for instance, the timelike line 1 and the spacelike line 3.[1] If we require, however, according to the relativistic light-geometry, that simultaneity is to be defined by

[1] It is even permissible to choose curved lines as the coordinate axes. The restriction of the time axis to a straight line signifies the restriction to a uniformly moving system in the sense of § 24, and the restriction of the space axis to a straight line means that ϵ in the definition of simultaneity (2, § 19) is constant and therefore independent of position and time.

$\epsilon = \frac{1}{2}$ (2, § 19) for every state of motion, then this requirement coordinates a definite space axis to every time axis. (In Fig. 31, for example, line 2 is coordinated to line 1.) It can be shown that this prescription corresponds in the geometrical representation to the requirement that the time axis and the space axis must form conjugate diameters of the hyperbolas.

We now look for a realization of the negative Δs^2, i.e., we look for a physical object that satisfies the relations of congruence defined by the hyperbolas of quadrants I and II. Let us turn to Fig. 32, in which we

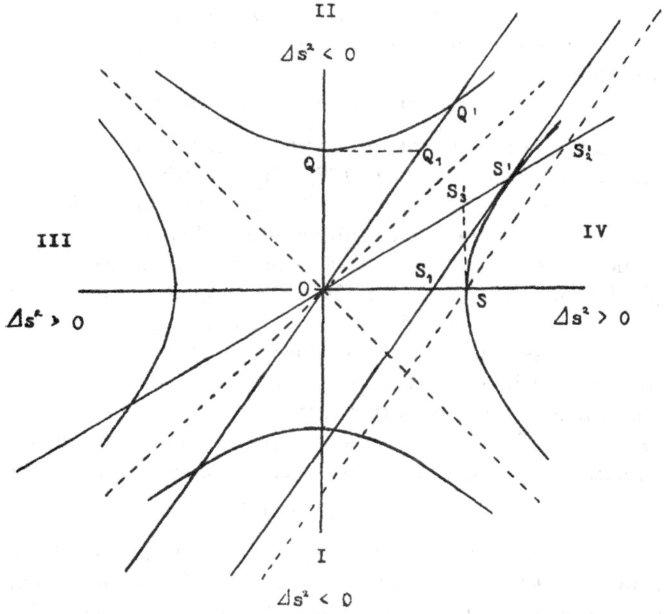

Fig. 32. Realization of the indefinite metric by means of clocks and measuring rods.

have drawn only the four branches of the hyperbola $\Delta s^2 = \pm 1$. Let us first consider the interval OQ, which belongs to the quadrant $\Delta s^2 < 0$. How can this interval be realized? The vertical axis OQ corresponds to the world-line of a point at rest in the coordinate system K. Events O and Q are therefore given by the beginning and end of a period of a unit clock at rest in K. If we write for these events the interval

$$\Delta x_1^2+\Delta x_2^2+\Delta x_3^2-\Delta x_4^2 = \Delta s^2 \qquad (1)$$

this form reduces to

$$-\Delta x_4^2 = \Delta s^2 \qquad (2)$$

because there is no change in position relative to K, i.e., $\Delta x_1 = \Delta x_2 = \Delta x_3 = 0$. The interval OQ is therefore measured by the period of a unit clock at rest in K.

Let us now consider the interval OQ', which equals OQ and also equals 1. It corresponds to the section OQ' of the world-line of the moving space point P' on which events take place at O and at Q'. Relative to K it is measured by (1). If we now change to a coordinate system K' by means of the Lorentz transformation, the same interval is expressed by

$$\Delta x'^2_1+\Delta x'^2_2+\Delta x'^2_3-\Delta x'^2_4 = \Delta s^2 \qquad (3)$$

since it is the peculiarity of the Lorentz transformation that it leaves this expression invariant (cf. p. 177). The coordinates determined by the light-geometrical metric are exactly those coördinates for which the metrical formula assumes this simple form. This result expresses the analogue of the geometrical assertion of the previous section that the simple formula of the metric (3*a*, § 28) applies only to rectilinear orthogonal coordinates. We can go even further. The coordinates given by measuring rods and clocks must correspond to those coordinates which satisfy (1) and (2) respectively, since this correspondence is expressed by the matter-axioms.

Now we can give a very simple interpretation of OQ'. If we choose a K' relative to which P' is at rest, then expression (3) reduces to

$$-\Delta x'^2_4 = \Delta s^2 \qquad (4)$$

for OQ', since we have again $\Delta x'_1 = \Delta x'_2 = \Delta x'_3 = 0$. This means that interval OQ' too is measured by the period of a unit clock, if this clock moves along the world-line OQ'. We therefore call Δs the *characteristic time* [1] of the clock and may now say: *a timelike interval is realized by the characteristic time of a moving clock, and the rotation of the interval OQ into the position OQ' is realized by placing the clock into a different state of motion.*

We shall now look for a realization of the positive Δs^2, i.e., for a physical object that satisfies the relations of congruence of the hyperbolas in quadrants III and IV. Let us first choose the interval OS

[1] I shall use the terms "characteristic time" and "characteristic length" as translations of the German terms "Eigenzeit" and "Eigenlänge."—M.R.

that corresponds to two simultaneous events in K, the spatial distance of which is equal to 1. Since now $\Delta x_4 = 0$, expression (1) reduces to

$$\Delta x_1^2 + \Delta x_2^2 + \Delta x_3^2 = \Delta s^2 \qquad (5)$$

This interval is therefore measured by the spatial length $\Delta\sigma^2$ of a unit measuring rod.

Now let us choose the interval OS', which equals OS and also equals 1. It is measured in K by formula (1). If we want to reduce (1) to an expression corresponding to (5), we must introduce a different simultaneity for which OS' is a simultaneity cross-section. This is easily accomplished if we place the rod into the moving system K', since OS' is the space axis corresponding to the time axis OQ' according to Einstein's simultaneity, if OQ' and OS' are conjugate diameters. The two ends of the rod describe the world-lines OQ' and S_1S', and the rod is represented by the slanted strip enclosed between the two world-lines. Therefore OS' is the position of the rod for a simultaneity cross-section x'_4; OS' is represented by the events occurring at its two ends, if these events are simultaneous in the sense of Einstein's definition of simultaneity for K'. This definition reduces (3) to the expression

$$\Delta x'^2_1 + \Delta x'^2_2 + \Delta x'^2_3 = \Delta s^2 \qquad (6)$$

Interval OS' is therefore likewise measured by the length of a measuring rod, if the rod is moved in the described fashion. Corresponding to the concept of characteristic time we now form the concept of *characteristic length*: the characteristic length of a measuring rod is determined by two events occurring at its endpoints, when these events are simultaneous according to Einstein's definition of simultaneity for that system in which the measuring rod is at rest. *A spacelike interval is realized by the characteristic length of a measuring rod, and the rotation of interval OS into position OS' is realized by placing the measuring rod into a different state of motion and by producing events at its ends that are simultaneous according to the appropriate definition of simultaneity.*

This result shows a peculiar difference between clocks and measuring rods. Clocks are inherently four-dimensional instruments, since the endpoints of their unit distances are *events*. Measuring rods, on the other hand, are three-dimensional measuring instruments; their endpoints are space points and they can be changed into four-dimensional measuring instruments only if events are produced at their endpoints according to a special rule.

These considerations show that clocks and measuring rods supply a realization of the indefinite geometry of Fig. 30*b*, and that this geometry characterizes the structure of the space-time manifold. *The assertion that measuring rods, clocks, and light rays behave according to the relations of congruence of the indefinite metric represents the geometrical formulation of the light- and matter-axioms.*

We have previously discussed (page 160) the assertion concerning the *union of space and time.* On the basis of the geometrical representation we can now clarify this assertion. Surely, the graphical representation of time, the combination of space and time into one manifold, is not new, since it also holds in the classical theory of time. The new content can be summarized in the following two assertions.

First, it is maintained that the element of the manifold determined by two point-events, namely the interval, finds its natural realization by clocks, measuring rods and light rays. This means that these measuring instruments introduce into the manifold certain congruence relations of a very specific kind. It is this fact which has made the four-dimensional treatment of space and time so fruitful, and which is expressed in the statement that clocks, measuring rods, and light rays assume for the four-dimensional space-time manifold a function which is similar to the function performed by rigid rods in three-dimensional space. It is true that the classical space-time theory could have treated space and time as a four-dimensional manifold; it would even have been possible to define some metric within this manifold. However, there would have existed no physical objects that would have realized the congruence relations of this metric. The assertion that there exists a natural metric for the space-time manifold has therefore great significance for physics. In this sense we may speak of a union of space and time. This does not mean, however, that space and time lose their specific individual differences, for, clearly, clocks and measuring rods are quite different types of measuring instruments. This union of space and time, therefore, preserves their specific properties.

The transition from the indefinite metric of the theory of relativity to the classical theory of time can be characterized by the replacement of the limiting velocity c by ∞. The basic metrical formula will then degenerate, since $x_4 = ct$ becomes infinite, and no metric of this kind is possible. If we were to construct a metric in the classical theory of time, we would not work with an indefinite metric, but would use the definite metric (3*b*, § 28) and define the congruence of four-dimensional intervals in an arbitrary fashion. Of course, there are such intervals even in the classical theory of space and time, but there is no natural rule, only arbitrary definitions, for their relations of congruence, i.e., for the measurement of length. The clock would not present a realization of these intervals, but would measure contour levels; see § 30.

Secondly, it is asserted that the manifold of space and time is of the indefinite type. This assertion has the consequence that not only the *time axis* can be rotated—this can also be done in the classical space-time theory, since it merely means that any moving system may be chosen as the one at rest—but even the *space axis* can be rotated. In the classical theory of space and time the space axis could not be rotated, because the angles of quadrants III and IV were equal to 0. These differences of the new space-time theory in this respect have been the cause of erroneous interpretations. It was believed that the *coupling* of the space and time axes supplied by the Lorentz transformation, according to which every choice of the time axis determines a corresponding space axis as the conjugate diameter, signifies a more fundamental junction of space and time. This coupling, however, is relatively unimportant because it is based on an arbitrary additional requirement, introduced only for descriptive simplicity, for which there is actually no epistemological need. The mistake committed here is the one pointed out on page 146; it springs from the erroneous conception that there is a relation between the relativity of simultaneity and the relativity of motion. Only the fact that the space axis can be rotated independently of the time axis is new in the indefinite metric. This statement means that simultaneity is arbitrary within a certain angular interval. We saw (§ 22) that this fact represents a specific feature of the relativistic theory of space and time, which we formulated as a structural property of causal chains and illustrated in Fig. 23. We showed in § 25 that the definitional nature of simultaneity results in an indeterminateness for spatial measurements; therefore we cannot speak of the state of a space at a definite time before a definition of simultaneity has been given. For this reason the indefinite metric formulates a second important point that can be referred to as a union of space and time. But this characterization of the space-time geometry can also be considered as leading to a sharp separation of space and time, because in the equation (1) of the indefinite metric one dimension is clearly distinguished from the others by the negative sign. It is true, of course, that even this representation does not exhaust the specific characteristics of time. This follows from the fact that the purely spatial metric of § 28 also represents the indefinite type, i.e., that an indefinite metric can likewise be realized by spatial measuring instruments alone. However, equation (1) expresses an asymmetry between space and time, which clearly distinguishes between the two.

The theory of relativity therefore does not maintain that time is the nsion of space"; it is and remains time in all its specific When we symbolized time by means of a straight line in 32, we only gave a *graphical representation*, which means that the logical structure exhibited by the rods described in § 28 can also be realized by the space-time manifold. This representation is analogous to the graphical representation of thermodynamic relations (§ 15), which exhibit the same logical structure as do rigid bodies (in this case the ordinary kind of rigid bodies). Therefore, the combination of space and time under one metric cannot erase their differences. If we speak of a geometrization of physical events, this phrase should not be understood in some mysterious sense; it refers to the identity of *types of structure* and not to the identity of the *coordinated physical elements*. On the contrary, there are essential properties of time which are not expressed in the geometrical representation in spite of the indefinite type of metric. However, if we remember these limitations, geometrical representation and visualization by means of drawings can always be used, because they combine the rigor of logical inference with mathematical elegance and lucidity.

§ 30. THE RETARDATION OF CLOCKS

In this and the next section we shall deal with two consequences of the matter-axioms which have given rise to misinterpretations and unjustified criticism. The first concerns clocks; the second, measuring rods.

Let us assume that two clocks U_1 and U_2 are set according to Einstein's definition (1, § 19) in an inertial system K (Fig. 33). A clock

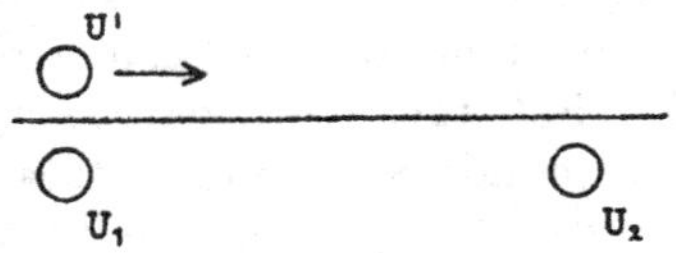

Fig. 33. Retardation of clocks.

U' with equal rest-unit (that is, if it were to remain at rest next to U_1 it would always show the same time as U_1) is moved in the direction of the arrow with a velocity v. When it is next to U_1 it shows the same time as U_1 (this determination requires only the neighborhood comparison of clocks); what will it show when it reaches clock U_2? Einstein's theory maintains that it will then be slow relative to U_2.

This statement is also called Einstein's retardation of clocks, according to which a moving clock is *slow* relative to the time of the rest-system.

The geometrical interpretation of this problem leads us back to the relations indicated in Fig. 19 (p. 134). Let us draw Fig. 19*b* again (Fig. 34) with the difference, however, that we do not use the moving

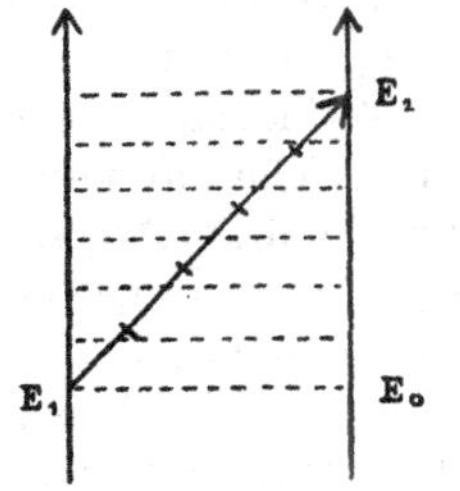

Fig. 34. The difference between interval and coordinate time.

clock in our definition of simultaneity, but compare it with a simultaneity obtained by Einstein's definition. The horizontal lines will now correspond to Einstein's simultaneity. Furthermore, we shall assume uniform motion, i.e., that the world-line E_1E_2 of the clock U' is a straight line. According to Einstein's retardation of clocks, the number of sections which the moving clock has marked off along its world-line E_1E_2 is different from the number of contour levels through which it has traveled.

The relations drawn in Fig. 32 (p. 185) supply a simple explanation. OQ is the period of a unit clock. The event Q_1 is simultaneous with Q, measured according to Einstein's simultaneity defined in K. When the moving clock reaches Q_1 it has not yet completed its period, since this is accomplished only at Q'. The difference between the classical and the relativistic theory of time, therefore, is expressed in our geometrical interpretation as follows: *according to the classical theory of time, a moving clock measures the coordinate time, and according to the relativistic theory it measures the interval.*

Which of the two theories of time is correct? At any rate, the distinction between interval and coordinate time is certainly correct. The difference between the two theories of time consists in a purely physical assertion. It is impossible to state *a priori* whether the moving clock indicates the coordinate time or the interval. Epistemologically speaking, both suppositions are possible; which of them applies to reality can be shown only by experience.

It is wrong, therefore, to say that the relativistic retardation of clocks is inconceivable. One phenomenon is as conceivable as the other, and we can only ask which one occurs in fact. Visualization cannot give us the answer; only observation can do so. The relativistic hypothesis has been confirmed by observation of traveling atom-clocks, which in fact show Einstein's retardation.[1] The classical theory is also based on a hypothesis, since it is a physical hypothesis that the clock which travels along E_1E_2 (Fig. 34) indicates the contour levels, and not the length of the world-line. This hypothesis is identical with the acceptance of a specific simultaneity, since the moving clock can indicate the time difference only for one system of contour lines, which it would thus distinguish from all others.

An objection to the relativistic theory of clocks that plays an important role in the literature, and may therefore be discussed, is given by the so-called *clock paradox*. The clock U' (Fig. 33) is slow compared to the time in K, and when it reaches U_2 will show an earlier time than U_2. Let us imagine that U' is stopped in its path at this instant and turned around so that it will travel back to U_1. The time taken in turning the clock around may be ignored, since it is negligible compared to the time of the round trip. The same retardation occurs during the return trip, and U' must therefore be slow compared to U_1 when it reaches U_1. The last statement is independent of both the simultaneity definition for K and the behavior of U_2. We can therefore say: if a clock U' is first moved away from U_1 and then returned, it will be slow relative to U_1.

It seems that according to the theory of relativity the process can be interpreted in the opposite manner. We consider U' to be at rest, while U_1 is moved (to the left) and then returned. On the basis of this description we should conclude that U_1 is slow relative to U', since U_1 is the moved clock. This result constitutes a contradiction, because a neighborhood comparison, independently of the definition of simultaneity, can tell us which of the clocks is slow when they meet. Only one of the two assertions can be correct.

The contradiction is quite striking and may under no circumstances be solved by considering the following two statements compatible: "when brought together U' is slow relative to U_1" and "when brought

[1] I refer here to the experiments of H. Ives on the transverse Doppler effect in cathode rays. For this question see also A., §§ 23–24.

together U_1 is slow relative to U'."[1] The comparison of the two clocks is independent of the definition of simultaneity. If the two above statements were both considered to be true, such a conception would

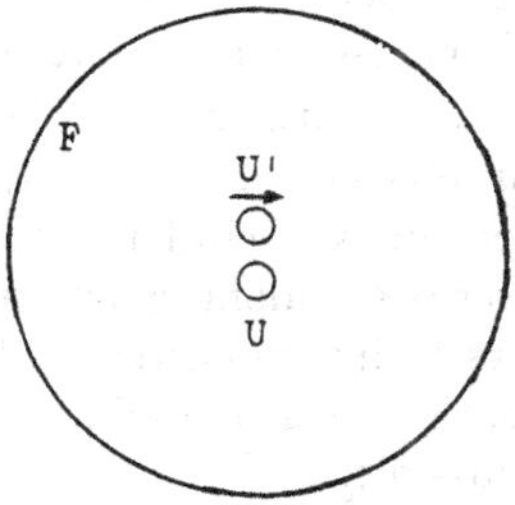

Fig. 35. The asymmetry in the clock paradox.

contradict the basic rule of the theory of relativity that a point-event (a coincidence) has an objective significance. A solution can be given only if we can show that one of the apparently equally correct inferences is incorrect. In fact, it is the second one.

The error lies in a misconception of relativity, which can be explained as follows. The theory of gravitation shows (cf. § 39) that the special theory of relativity is applicable only because the distant masses of the fixed stars (drawn as a circle in Fig. 35) determine a particular metrical field. If we take account of the masses of the fixed stars F, the apparent equivalence of the two interpretations vanishes. According to the first interpretation U' is moving, while U and the fixed stars F remain at rest. According to the second interpretation U' is at rest, and U and the fixed stars F are in motion. This analysis eliminates the symmetry of the two processes; the second is an entirely different process from the first because of the effect of the moving fixed stars, which produce a gravitational field at the instant of the reversal of the motion and thus cause a retardation of U'. Due to the gravitational field U' is the retarded clock even according to the second interpretation. Calculations prove this conclusion quantitatively correct.[2] The mistake that led to the paradox therefore resulted from the fact that the considerable effects of gravitation were ignored.

[1] This is the opinion of J. Petzold, *Die Stellung der Relativitätstheorie in der geistigen Entwicklung der Menschheit,* Dresden 1921, page 104.

[2] Cf., e.g., A. Kopff, *Grundzüge der Einsteinschen Relativitätstheorie,* Leipzig 1921, pages 117 and 189.

A remark may be added concerning the extension of Einstein's theory of clocks to living organisms. The retardation of clocks has often been illustrated by the example of the twins: of two newborn twins, one makes a cosmic trip with a velocity slightly below the velocity of light and returns as a boy, while the other twin has in the meantime become an old man. This consequence, which many people have regarded as absurd, actually contains nothing impossible or inconceivable and agrees with the theory of relativity in every respect. In fact, a similar case has been described in W. Müller's poem "Der Mönch von Heisterbach," which describes a monk going on a walk and returning after three hundred years to his monastery, where nobody recognizes him. The poet's imagination thus created ideas which modern physics no longer regards as impossible.

If the objection is raised that the theory of relativity as a physical theory applies to physical processes alone and not to living organisms, one forgets that there are many basic principles of physics which also apply directly to living beings. Galileo's law of falling bodies governs a falling stone as well as a falling egg or a falling human being. The laws of gravitation apply in general to animate and inanimate objects in equal fashion. After the discovery of the spherical shape of the earth, it was immediately inferred from this physical theory that human beings who live on opposite sides of the spherical surface nonetheless have the subjective feeling of upright posture, because living organisms adjust themselves to the physical gravitational field. It is a similar claim which is made by the theory of relativity in the example of the twins, namely, that living organisms, like clocks, adjust themselves to the metrical field. It would be an unjustifiable hypothesis to assume that they would behave differently, since the principle that the time scale of natural clocks is identical with the time scale of living organisms (insofar as it can be defined) is one of the oldest principles of natural science. The example of the twins is explained by the fact that the ultimate constituents of living organisms are atoms. If every atomic period, i.e., the period of the electron within the atom, is retarded to the same degree under the influence of motion or of a metrical field, physiological phenomena would show the same retardation. since they result from the integration of many atomic periods.

§ 31. THE LORENTZ CONTRACTION AND THE EINSTEIN CONTRACTION

The theory of relativity makes an assertion about the behavior of rigid rods similar to that about the behavior of clocks. This assertion states that the characteristic length of a rod measures the interval. On the basis of the geometrical interpretation of Fig. 32 (page 185) we can easily recognize that this statement leads to consequences different from those of classical space-time theory. According to classical theory, the moving rod is not represented by the strip between the world-lines OQ' and S_1S', but by the wider strip bounded by OQ' at the left and SS'_2 at the right. This follows, according to classical theory, because the length of the moving rod, measured in K, is given by OS, and its rest-length in K' would correspondingly be OS'_2. According to the theory of relativity, the rod has the shorter rest-length OS' in K'.

This assertion of the theory of relativity is based mainly on the Michelson experiment.[1] This experiment proves that the rods satisfy the light-geometrical definition of congruence (cf. Fig. 29, page 170) where

$$AB = AC \text{ when } \overline{ABA} = \overline{ACA} \qquad (1)$$

in every inertial system, for any orientation of the coordinate axes. According to classical theory, (1) is satisfied in only *one* of the inertial systems, namely, the system at rest relative to the ether. In all other systems the rest-length of one of the branches of the coordinate axes will no longer satisfy (1). Since the Michelson experiment has been confirmed to a very high degree, we could consider this matter closed, because it has no bearing upon epistemological considerations, if it had not been given certain erroneous interpretations in the usual discussions on relativity.

In order to explain the experiment Lorentz made the assumption that one arm of the apparatus is contracted by the amount $\sqrt{1-\frac{v^2}{c^2}}$ when it moves relative to the ether. Einstein, on the other hand, considered both arms equally long in every inertial system and calculated the contraction factor $\sqrt{1-\frac{v^2}{c^2}}$ in an entirely different manner, as a consequence of the relativity of simultaneity. The opinion has been expressed that the contraction of one arm of the

[1] It does not follow, of course, from this experiment alone. See A., §§ 21, 24.

apparatus is an "ad hoc hypothesis," while Einstein's hypothesis is a natural explanation that is a consequence of the relativity of simultaneity. Both of these explanations are wrong. The relativity of simultaneity has nothing to do with the contraction in Michelson's experiment, and Einstein's theory explains the experiment as little as does that of Lorentz.

The above opinion is incorrect because the contraction of the arm of the apparatus occurs for the moving system relative to which the apparatus is at rest. The Einstein contraction would explain a contraction of the arm only if it were measured from a different system and is therefore not sufficient to explain the Michelson experiment. This experiment proves that a rod which lies along the direction of the motion is shorter than it should be according to classical theory, *if it is measured relative to the rest-system.* In other words: the comparison of the rest-length of moving rods does not obey classical theory. If there were a special inertial system I that could be regarded as an absolute rest-system, and if we had in this system two equally long rigid rods, one of which behaved according to classical theory and the other according to Einstein's theory, the two rods would cease to be equally long if they were brought into any other inertial system S, provided that they lay along the direction of the motion of S. The Einstein rod would be shorter. The difference could be measured in S as the difference between their rest-lengths, and in any other system as the difference between the lengths of the moving rods. Einstein's theory as well as Lorentz's theory therefore assumes the behavior of rigid rods to be measurably different from their behavior according to classical theory; but the difference has nothing to do with the definition of simultaneity.

It has been objected to previous remarks of mine on this subject [1] that it is impossible to compare two magnitudes belonging to different theories. This objection is incorrect. By reference to a third body, we are able to establish a comparison, if we calculate how the two magnitudes under consideration compare with that third body. Furthermore, this mode of expression is frequently used in physics. We may say, for instance, that a highly-compressed gas behaves differently than it would according to Mariotte's law. This means nothing but that the real gas g, when compressed to a certain degree, occupies a larger volume than a gas G which satisfies the Mariotte-Boyle relations. The third body used in this comparison is the rigid

[1] See *Zs. f. Phys.* 34, 1925, pp. 44f.

measuring rod which measures the volume. The third body is not always explicitly mentioned, and the abbreviated formulation is often preferred, because it clearly suggests a difference in actual behavior. The Lorentz contraction must indeed be considered a *real difference* in this sense. In this case, the *tertium comparationis* is light, which in terms of light-geometrical definitions supplies a standard to which the rods of the different theories can be compared. It would be an incorrect mode of speech, however, to say that the Einstein contraction is an *apparent difference.* This contraction has nothing to do with the difference between the real and the apparent, but results from a *difference in the conditions of measurement.* We shall speak of a *metrogenic* difference because this difference originates in the nature of the measurement. Since we are specifically concerned with kinematic conditions of measurement, we shall speak of a *metrokinematic difference.*

We have pointed out before that the length of a rod is not the rod itself, but a logical function thereof; it is a number coordinated to the rod expressing its relation to other rods. The process of coordination depends on certain specified conditions. In the discussion of the concept of the length of a moving rod (§ 25) we emphasized that the coordinated length l depends on the simultaneity s as well as the velocity v. It is to be expected, therefore, that the coordinated length l will vary with changes in the two parameters s and v. This distinction is inherent in the kinematic system. Just as we spoke of perspective differences, in the previous example about the angle of vision, we may now speak of metrokinematic differences. Without changes in the rod itself, it is subjected to different kinematic conditions of measurement and will therefore yield different measures within the frame of one consistent theory.

The situation is different for the Lorentz contraction. It compares rods under *the same conditions of measurement* relative to *different theories.* Here we have conflicting empirical assertions, since the two theories exclude each other. The same rigid rod behaves differently under the same conditions of measurement whether the Lorentz or Einstein theories on the one hand, or the classical theory on the other hand, are correct. We therefore speak of a *real difference* when we compare the real behavior with a possible behavior of objects. This real difference exists, in the explanation of the Michelson experiment, between the classical theory and Einstein's theory as well as between the classical theory and Lorentz's theory, whereas there is no difference between Einstein's and Lorentz's theories; both assert the state of affairs formulated in (1), while the classical theory differs from them in this respect. The concept of simultaneity does not enter into this problem at all.

It would be advisable, therefore, not to use the same name for the two "contractions." There is an *Einstein contraction*, which results from the relativity of simultaneity and compares the length of the moving rod with the length of the rod at rest; and there is a *Lorentz contraction*, which compares the length of a rigid rod that satisfies the Michelson experiment with the length of the rod as defined in the classical theory. It is a coincidence that both have the same contraction factor $\sqrt{1-\frac{v^2}{c^2}}$ and this is probably the reason that the two contractions have been so frequently confused. Their meanings are different. In addition to the Einstein contraction, Einstein's theory also contains the Lorentz contraction, which it "explains" as little as does the Lorentz theory. It is simply assumed axiomatically.

What is the difference between Einstein's and Lorentz's theories? In order to answer this question let us distinguish between the following two statements:

(a) the rest-length of the moving rod is different from the rest-length of the rod at rest.

(b) the rest-length of the moving rod is different from the rest-length of another rod which moves with it but satisfies the classical theory.

Statement (b) is true. This was proved by the Michelson experiment and some further assumptions; we shall call it the Lorentz contraction. In the geometrical representation it is indicated by the difference between the distances OS' and OS'_2 (Fig. 32, p. 185). As formulated, statement (a), on the other hand, is neither true nor false; its truth depends on the *coordinative definition for the comparison of the rest-lengths of moving segments* (cf. § 25). In the geometrical representation, statement (a) is equivalent to the comparison of OS' and OS. It seems that Lorentz believed that statement (a) follows from statement (b). This belief, however, would constitute an epistemological error. Einstein's theory rejects Lorentz's conclusion and, recognizing the existence of a coordinative definition, regards the two rest-lengths mentioned in (a) as equal. It is sometimes overlooked by proponents of the theory of relativity that statement (b) is nevertheless true. It follows that Einstein's theory also contains a contraction which is independent of the relativity of simultaneity, namely, the Lorentz contraction. In addition, however, it contains the difference between the rest-length and the length of a moving segment, in other words, the Einstein contraction.

If we say that the equality of the two contractions is "coincidental," this means that their equality depends on certain presuppositions; nevertheless there is a theoretical connection between them. It can be shown that they must always be equal if and only if the transformation is linear. The proof is as follows. Let l be a rod that behaves according to the Lorentz or Einstein theory and L be a rod that behaves according to classical theory. Their rest-lengths in K are equal, or $l_K^K = L_K^K$ (where the upper index designates the system in which the measurement is performed, and the lower index the system in which the rod rests). The Lorentz contraction therefore concerns the ratio

$$l_{K'}^{K'} : L_{K'}^{K'} \tag{2}$$

while the Einstein contraction concerns the ratio

$$l_{K'}^{K} : l_K^K \tag{3}$$

According to classical theory we have $L_{K'}^K = L_K^K$ (this comparison uses the simultaneity of K and is independent of that of K'), and since $l_K^K = L_K^K$ we obtain $L_{K'}^K = l_K^K$. Relation (3) therefore becomes

$$l_{K'}^{K} : L_{K'}^{K} \tag{4}$$

Because of the linearity of the transformation (and only because of it), ratio (4) is the same as ratio (2), which means that (3) is also the same as ratio (2).

On the other hand, the ratio

$$l_K^{K'} : l_{K'}^{K'} \tag{5}$$

which equals (3) according to the Lorentz transformation, can be quite different even if we use linear transformations. This fact permits us to construct an example in which an Einstein contraction appears but no Lorentz contraction. If the rigid bodies did not behave like l in the relativistic sense, but like L in the classical sense, there would be no Lorentz contraction; however, the Einstein contraction from K' to K would also disappear, since

$$L_{K'}^{K} : L_K^K = 1 \tag{6}$$

If we nevertheless define the simultaneity in K' according to Einstein, setting $\epsilon = \frac{1}{2}$ (see 2, § 19), the inverse comparison from K to K' will show the Einstein contraction

$$L_K^{K'} : L_{K'}^{K'} = 1 - \frac{v^2}{c^2} \tag{7}$$

the magnitude of which is the square of that of the Lorentz-Einstein contraction. This proof clarifies the fundamental difference between the two contractions.

The difference is perhaps most conspicuous in the geometrical representation of Fig. 32 (p. 185). The Lorentz transformation claims that OS' is shorter than OS'_2, since $OS' = l_{K'}^{K'}$ and $OS'_2 = L_{K'}^{K'}$. The Einstein contraction maintains that OS_1 is shorter than OS, since $OS_1 = l_K^K$ and $OS = l_{K'}^K$. It can easily be seen that the second statement is a consequence of the first, because, according to the relativistic theory of time, the rod is represented by a narrower strip. Even if the classical theory were correct, and the rod were represented by the wider strip, $OS'_3 = L_K^K$ would still be shorter than $OS'_2 = L_{K'}^{K'}$ (SS'_3 is parallel to OQ and these two lines are the boundaries of the world-strip of the rod at rest in K). This means that for K' there is an Einstein contraction but no Lorentz contraction.

Another example of an Einstein contraction without a Lorentz contraction results if the rods behave like L, i.e., there is no Lorentz contraction, but simultaneity in K is defined differently from Einstein's $\epsilon = \frac{1}{2}$. This makes $L_{K'}^K : L_K^K \neq 1$. The example used in a similar form in § 25, where we discussed the dependence of the length of the moving segment on the definition of simultaneity, makes it particularly clear that the Einstein contraction is a metrogenic phenomenon. In the geometrical representation this means that we may choose as the length of the rod differently directed sections through the world-strip of the rod. On the other hand, the geometrical representation of Fig. 32 shows very clearly that through the difference in the width of the strip, the Lorentz contraction indicates a difference in the actual behavior of the rod. These considerations also explain how it is possible to compare rods l and L, although only one of them is physically realized. OS is the same in both theories; the classical theory claims that the right-hand boundary of the strip parallel to OQ' must be drawn through S, whereas the new theory places the boundary along the tangent to the hyperbola which passes through S.

If we wish to enter into the further question of an explanation, we should first point out that the problem has been confused by the use of the word "contraction." This word has led to a mistaken application of the principle of causality. Looking for a *cause* of the contraction, some philosophers and physicists have believed that a cause for the *difference* between the magnitudes must be found. This conception assigns a preferred position to one theory, namely, the classical theory, whose laws are taken to be satisfied by physical objects without a cause, while causes are made responsible only for deviations from this behavior. The problem of causality is to be posed in a different form; if we ask for a causal explanation, we must ask why it is that measuring rods and clocks adjust themselves to a certain transformation defined in terms of light-geometry. This causal problem is the same whether the rods and clocks adjust themselves to the relativistic or to the classical transformation. The word "adjustment," which was first used by Weyl in this connection (cf. § 39), characterizes the problem very well. It cannot be an accident that two measuring rods which have the same length at one place always have the same length when brought to a different place along different paths. It must be explained as an *adjustment to the field* in which the measuring rods are embedded like test-bodies. Just as the compass needle adjusts its direction to the magnetic field of its immediate environment, so measuring rods and clocks adjust their *unit length* to the metrical field. All metrical relations between physical structures must be explained in this fashion, including the Michelson experiment, according to which rigid rods adjust themselves in a definite way to the motion of light. The answer can of course be given only by a detailed theory of matter, which has not yet been elaborated. It must be explained why the accumulations of density in certain parts of the field, electrons and similar particles, provide a simple indication for the metric of the surrounding field. The word "adjustment" therefore poses a problem rather than supplies a solution. The existing situation is formulated rigorously by the matter-axioms without the use of the word "adjustment." If this theory of matter were exactly formulated, we would be able to explain the metrical behavior of physical structures. For the time being, however, we can speak of an explanation by Einstein's theory as little as we can speak of an explanation by Lorentz's theory or by the classical theory.

Why is Einstein's theory better than Lorentz's theory? It would be mistaken to argue that Einstein's theory gives an explanation of Michelson's experiment, since it does not do so. Michelson's experiment is simply taken over as an axiom. However, Einstein's theory is superior to Lorentz's theory because it renounces the quest for an explanation of the Michelson experiment in terms of a "contraction." The "explanation" given by Lorentz's theory is its weakness. It assumes the classical relations to be "self-evident" and postulates incorrectly that any deviation from these relations must have a cause. Einstein's theory makes use of the arbitrariness of the *coordinative definition for the comparison of the rest-lengths of moving segments* and calls the two rods equal if they behave in accordance with the Michelson experiment. The superiority of Einstein's theory lies in the recognition of the epistemological legitimacy of this procedure.

§ 32. THE PRINCIPLE OF THE CONSTANCY OF THE VELOCITY OF LIGHT

Finally, we must mention in this connection the constancy of the velocity of light, because this is claimed to be another assertion of the theory of relativity which is impossible to visualize. The statement that the same light impulse can be considered as a spherical wave relative to systems in different states of motion is only an apparent contradiction. It is solved if we remember that the wave surface is not a material surface, but an ideal structure projected into a process of propagation. Its shape therefore depends on the nature of the construction, which is determined by the method of the measurement of length, hence the definition of spatial congruence, and by the definition of simultaneity. Since we are concerned with the shape of a *moving* structure, measured from a *rest-system,* its shape can only be considered as its simultaneity projection on the rest-system. If the definition of simultaneity is changed, the shape of the projection is changed, and if the Lorentz transformation assumes a specific definition of simultaneity for every state of motion, then the shape of the simultaneity projection can easily be made spherical for every state of motion. If it is argued that it is impossible to visualize the same light impulse in the shape of a spherical wave for each of two systems in different states of motion, we answer that this apparent impossibility results from the tacit introduction of assumptions ignoring the demands of the theory. In fact, the theory does not say that the wave surfaces in the two systems consist of the same point-events; rather, every wave surface compared to that of the other system represents a "focal-plane shutter photograph" of the motion of light.

We may recall § 26, where we have shown that every centro-symmetric process of propagation, not only light, is an indefinite structure, and that by itself it does not define the state of motion of its center; this is accomplished only by the addition of a definition of simultaneity. There we showed that the propagation of the same light impulse can be conceived either as a system of concentric spheres or as a system of eccentric spheres, the centers of which move at a uniform velocity. The definition of simultaneity used in § 26 was not that of the Lorentz transformation, because we defined simultaneity from a system relative to which the centers were in motion. If the definition of simultaneity is given from a moving system, the spherical surface will result when Einstein's definition with $\epsilon = \frac{1}{2}$ is used, since it is this definition which makes the velocity of light equal in all directions.

The core of the principle of the constancy of the velocity of light, therefore, is not the assertion that light may be conceived as a spherical wave for every moving system. This part of the principle expresses only a general property of all centro-symmetric processes of propagation. Rather, the assertion that light possesses unique properties constitutes the physical content of the principle. We should therefore speak of the *principle of the uniqueness of light.* This principle has two parts.

The first part deals with the *principle of the limiting character of the velocity of light.* According to this principle, light has the highest velocity and is identical with the first-signal defined in § 22. This statement, which belongs to the light-axioms, is not an arbitrary assumption but a physical law based on experience. In making this statement, physics does not commit the fallacy of regarding absence of knowledge as evidence for knowledge to the contrary. It is not absence of knowledge of faster signals, but positive experience which has taught us that the velocity of light cannot be exceeded. For all physical processes, the velocity of light has the property of an infinite velocity. In order to accelerate a body to the velocity of light, an infinite supply of energy would be required, and it is therefore physically impossible for any object to attain this speed. This result was confirmed by measurements performed on electrons. The kinetic energy of a mass point grows more rapidly than the square of its velocity, and would become infinite for the speed of light. We cannot say that this conclusion is logically necessary. Whether or not physically possible velocities have a finite upper limit can only be taught by experience. If observation shows the existence of a limit, we have to accept it, instead of hoping that some day a greater velocity may be found. It is true that science has discovered many things which were never even thought of before, but these were mostly new discoveries that did not contradict previously established scientific laws. Of course, such a refutation may occur and then we would have to admit that the previous scientific "law" was false. This possibility should not prevent us, however, from believing in scientific laws so long as they are confirmed by experience.

Incidentally, the limiting character of the velocity of light can be made plausible in a very simple fashion. Light represents only a small section of the infinite range of electromagnetic waves, all of which have the same velocity. According to the present state of scientific theory, the electromagnetic wave is the *archetype of causal propagation*. Every other causal process, e.g., the propagation of elastic forces in a rigid body, or the flow of an electric current, is reducible to elementary electromagnetic processes.[1] It seems plausible that a causal process composed of many such elementary processes should be slower than the electromagnetic propagation; for instance, the collisions of the individual electrons may cause a retardation; however, it would be inconceivable that such a compound causal propagation would be faster. It follows that present-day physics has important positive grounds for asserting that the velocity of light is the *limit of the velocity of all causal propagation.*

The second part of the light principle can be formulated as the *principle of the metrical uniqueness of light.* It contains the assertion that the natural geometry of light rays, the light-geometry, is at once the geometry of rigid rods and clocks. This statement accounts for Einstein's term "the principle of the constancy of the velocity of light." It states that the velocity of light is constant even if the space-time metric is defined by means of rods and clocks. Only this formulation makes the principle complete and clarifies its physical import. It asserts a correspondence between light- and matter-geometry, a formulation that makes the empirical nature of the principle obvious. Einstein's terminology, however, may give rise to misunderstanding. Physics cannot assert that the velocity of light *is*

[1] An exception is the propagation of gravitional forces, which however according to Einstein also travel with the speed of light.

constant, since the measure of velocity contains an arbitrary element in the definition of simultaneity. We can only say that the velocity of light *can be defined* as constant without leading to contradictions. The "can," however, contains a far-reaching assertion about physical reality, namely, the identity of light-geometry and matter-geometry. To speak of the principle of the metrical uniqueness of light therefore seems to describe the situation more adequately.

Occasionally, the objection has been raised against the relativity of simultaneity that Einstein's definition could just as well have been given by the use of some other signal, such as sound; this is true. The use of any signal is justified by the complete relativity of simultaneity, i.e., the possibility of choosing for ϵ any value between 0 and 1. We should not believe, however, that this *sound-geometry* has the same unique characteristics as the light-geometry. The sound-geometry would not be identical with the geometry of clocks and rods, nor would it exclude velocities above that of sound. Because of the latter fact, simultaneity as defined in this case would lead to contradictions within the causal definition of time order as soon as signal velocities of a speed higher than sound were used. The special properties of light are not *consequences* of the theory of relativity but its *presuppositions*. Conversely, these empirical facts make the relativistic space-time theory physically significant, for they make the light-geometry the natural geometry of physics.

E. Cohn [1] has constructed a very instructive model of the Lorentz transformation, which reproduces all its relations correctly and exhibits very clearly the relativity of simultaneity. The model presents the velocity of light on a reduced scale, and therefore units are chosen for the two moving systems which do not coincide if transported into the same state of motion. These adjustments increase the illustrative power of the model, since it demonstrates the uniqueness of light in a very effective manner. The epistemological objections raised by J. Petzold [2] against this model cannot be maintained.

The principle of the uniqueness of light is one of the most basic assumptions of the theory of relativity. Its character as a physical and not as an epistemological assumption must be recognized before the space-time doctrine of the theory of relativity can be judged. Its division into two parts is also important for the physical theory, since it is only the first, *the principle of the limiting character of the velocity of light*, which is taken over into the general theory of relativity. The second part, namely the *principle of the metrical uniqueness of light*, holds

[1] E. Cohn, *Physikalisches über Raum und Zeit*, Teubner 1910 and later ed.
[2] J. Petzold, *Verh. d.d. phys. Ges.* 1919, p. 495.

only in the special theory of relativity, while it is abandoned step by step in the general theory; this will be shown in § 38 and §§ 41–42.

§ 33. THE ADDITION THEOREM OF VELOCITIES

Finally let us investigate the problem of the addition of velocities, which is solved in the theory of relativity by means of the peculiar addition theorem of Einstein. If a body has two velocities at the same time, how can these be combined into one resultant velocity? This question is significant because it also concerns the problem of resolving a velocity into its components; the combination of the components of a velocity into a resultant is only a special case of the addition theorem in which the individual velocities are perpendicular to each other. We must now investigate to what extent the addition of velocities can be derived logically and to what extent it is a matter of experience.

First let us state the problem more precisely. What does it mean to say that a body has two velocities at the same time? Obviously this expression is very vague. We may say, for example: a body has at one time the velocity $\vec{u}$ and at another time the velocity $\vec{v}$; what velocity $\vec{w}$ results if it has both velocities at the same time? This question leads back to the first question, since we are using the phrase "having at the same time." It is therefore preferable to reverse the question: a body has a velocity $\vec{w}$, and we ask how this velocity can be resolved into two components.

We may proceed as follows, provided that we consider only uniform rectilinear motion; all other motions are reducible to this in infinitesimal regions. Let a body move from P_0 to P_1 with a velocity $\vec{u}$ and then from P_1 to P_2 with a velocity $\vec{v}$ in a different direction. For each of these paths it uses the same time interval Δt. If we send it another time directly from P_0 to P_2 with a velocity of $\vec{w}$ such that it takes the same interval Δt, which it had previously used for the separate parts of the trip, we say that it has velocities $\vec{u}$ and $\vec{v}$ at the same time. Now we have given a definition of "having at the same time," and we can derive logically the addition theorem for vectors

$$\vec{u}+\vec{v}=\vec{w} \tag{1}$$

In this form the addition theorem is a logical consequence; and at the same time it justifies the resolution of a velocity into its components.

The components of the velocity are defined in terms of the total velocity in the same way as the individual velocities $\vec{u}$ and $\vec{v}$ were defined in terms of $\vec{w}$. The law of the components of a velocity, which is a special case of (1), is therefore purely a matter of logic.

We can also give the expression "having at the same time" a completely different meaning which has a direct physical interpretation. For this purpose we use an auxiliary system K'; the system of reference is K. Let K' have the velocity $\vec{u}$ relative to K and let a given body have the velocity $\vec{v}$ relative to K'. What velocity $\vec{w}$ does this body have relative to K? More precisely, how is its velocity $\vec{w}$ relative to K computed from $\vec{u}$ and $\vec{v}$?

It is immediately obvious that an answer to this question will depend on the definitions of the measurements of length and of simultaneity in the two systems K and K'. Let us assume that these two measurements are defined in a certain manner in K', and therefore, that the body has the velocity $\vec{v}$ relative to K'. Its velocity relative to K will then be $\vec{w}$. If we change the definitions for the two measurements in K' and consider a body which now has the velocity $\vec{v}$ relative to K', this body will have a state of motion different from that of the first and it will therefore have a different velocity $\vec{w}^*$ relative to K. Consequently its velocity $\vec{w}^*$ must be calculated differently, in terms of $\vec{u}$ and $\vec{v}$, from the velocity $\vec{w}$ of the first body. We can therefore give an addition theorem corresponding to the second meaning of "having at the same time" only if we state how length and time measurements are to be performed in K and K'.

It can easily be shown how these measurements have to be performed in order to obtain addition theorem (1). For this result we must require:

(1) Simultaneity and the measure of time in K' are to be defined so that every clock in K' always shows the same time as that clock in K which it is passing at that moment. In other words: the time of K' is identical with the time of K.

(2) The length of a distance l' in K' is to be measured by that distance l of K which is the projection of l' on K on the basis of the simultaneity of K. In other words: the measurement of length in K' is identical with that of K.

It can easily be seen that the vector-addition theorem (1) follows logically from these conditions, since they define the so-called Galilei-transformation

$$x_\alpha = x'_\alpha + u_\alpha t \qquad \alpha = 1, 2, 3 \qquad t = t' \tag{2}$$

in which the components u_α of $\vec{u}$ are defined according to the above rules.[1]

The problem is essentially different for the relativistic theory of space and time. It defines the measure of time, simultaneity, and the measurement of length in K' in a different fashion, requiring that the relativistic light-geometry be established in K' as well as in K. Since this requirement is equivalent to the use of the Lorentz transformation for the transition from one system to the other, we obtain a different addition theorem for the relativistic light-geometry. This theorem is derivable from the Lorentz transformation; and we shall present it here for the simple case where $\vec{u}$ and $\vec{v}$ are equidirected:[2]

$$w = \frac{u+v}{1+\frac{uv}{c^2}} \tag{3}$$

Like additional theorem (1), this theorem is logically necessary, but it rests on different assumptions.

We may also consider this addition theorem as an empirical result, if we define the metric in K' not light-geometrically, but with the aid of clocks and rigid rods. Then (3) states: if lengths and time intervals are measured in K and K' by the same kind of natural clocks and rods, and if simultaneity is defined for both systems according to Einstein's formula (1, § 19), then addition theorem (3) holds for the second kind of definition of "having at the same time" two velocities. This is obviously an empirical law, since it concerns clocks and rods the behavior of which cannot be determined *a priori*. On the other hand, it is clear that this law has no other empirical content than that expressed by the correspondence between light-geometry and matter-geometry. With this formulation we have exhausted the empirical content of the relativistic kinematics as far as concerns measuring instruments.

[1] See A., §§ 14–15, where the Lorentz transformation is derived via the Galilei-transformation and where it is explained that the difference between the two transformations is merely definitional.

[2] For the general case see M. v. Laue, *Relativitäts-Prinzip*, Vol. I. Braunschweig 1913, p. 46.

Finally, retaining the second meaning of "having at the same time" two velocities, we can give the addition theorem another interpretation by using the following rule for the measure of velocity in K'. Let us erect in K a mechanism that would give a body the velocity $\vec{v}$ if it were fired in K, e.g., a cannon with a certain powder charge. If this cannon is transported to K' and fired, then we say that the shell has the velocity $\vec{v}$ relative to K'. Since K' moves with the velocity $\vec{u}$ relative to K, how can $\vec{w}$ be calculated from $\vec{u}$ and $\vec{v}$? The velocity in K' is not defined by clocks and rods, but by the transport of a firing-mechanism. Therefore the question arises: to what definition of the space-time metric in K' does this arrangement correspond?

Relativistic mechanics claims that this arrangement corresponds to the relativistic definitions for measurements in K', while classical mechanics claims that it agrees with those definitions for measurements in K' determined by conditions (1) and (2) above (p. 207). These are two entirely different assertions about the world. The theory of relativity states that under the same physical conditions in K' the same velocity will be produced relative to the relativistic metric, while the classical theory maintains that the corresponding velocity is the same only if the measurements in K' are performed according to rules (1) and (2). For the theory of relativity, the relativistic metric of K' is the normal system.

This analysis supplies additional empirical content for (3); the composition of velocities that are produced by a series of physical mechanisms satisfies the addition theorem (3). For this reason a repeated operation of firing mechanisms will never produce the velocity of light, although each firing from an already moving mechanism will increase the velocity. The velocities will approach the velocity of light asymptotically; this result can be inferred from (3) by simple calculation.[1]

The last assertion goes beyond relativistic statements about space and time in the narrower sense, and leads into relativistic physics; the principle of relativity therefore contains statements not only about space-time measuring instruments but also about all physical phenomena in general. The pursuit of this aspect of the theory of relativity does not lie within the scope of this book, which is restricted to the problems of space and time.

[1] It can easily be shown that if $u < c$ and $v < c$, then also $w < c$.

B. Gravitation-Filled Space-Time Manifolds

§ 34. THE RELATIVITY OF MOTION

Considering now the problem of motion, we turn to a treatment of the problem of space which is historically speaking much older than the geometrical treatment of the problem. Whereas the latter approach developed only after the invention of non-Euclidean geometry in the beginning of the 19th century, the relativity of motion had already attracted attention at the time of Newton and Leibniz. Even at that time there was a dispute between relativists and antirelativists, and the well-known correspondence [1] between Leibniz and Clarke, a follower of Newton, became the talk of the day. Whoever reads those letters today finds in them many of the arguments and objections which are known from modern discussions of the problem of motion.

That the relativity of motion was defended at such an early date is due to its conspicuousness. Motion is change in position; it is clear, however, that it cannot be observed unless it is a change in position relative to a certain body and not relative to an ideal space point. Is it meaningful, under these circumstances, to speak of absolute motion or of motion relative to space, if motion relative to other bodies only can be observed? The distinction between what is observable and what exists seems reasonable at first, but becomes very doubtful on closer inspection; there is a strong feeling that it is meaningless to postulate differences in objective existence if they do not correspond to differences in observable phenomena. Leibniz expressed this idea in his *principle of the identity of indiscernibles*, from which he derived a theory of the relativity of motion, which even today forms the basis of the theory of relativity. According to this principle there exists only a motion of bodies relative to other bodies, and it is impossible to distinguish one of these bodies as being at rest, because rest means nothing but rest relative to another body, i.e., rest is itself a relative concept. We shall call this conception *kinematic relativity*. Motion as a kinematic process, as a change in spatial distances, is relative, since all kinematically observable phenomena are the same whether one or the other of two bodies is assumed to be at rest. The cosmologies of

[1] See A. Buchenau and Ernst Cassirer, *G. W. Leibniz Philosophische Werke* (4 vols.): *Hauptschriften zur Grundlegung der Philosophie* (2 vols., Leipzig, 1924) vol. I, p. 120 f.

Copernicus and Ptolemy are kinematically equivalent; both of them are descriptions of the same facts, and Ptolemy's epicycles of the planets are the kinematic equivalents of the circular orbits of Copernicus.

Even this early discussion of the problem of motion, however, deals already with another kind of question, namely the *dynamic problem of motion.* It was Newton, the father of mechanics, who introduced criteria of motion other than kinematic ones into the discussion. Discovering the quantitative relation between motion-producing forces and a kinematic magnitude, the acceleration, he could in turn use force as a measure of acceleration, i.e., the state of motion. Newton therefore denied the relativity of motion. According to him it is not true that all observable phenomena are the same regardless whether we consider one or the other of two bodies as being at rest, because differences appear as soon as dynamic phenomena are included in the observations. The measurement of the effect of a force therefore permits the determination of a state of motion relative to space. This leads Newton to the concept of absolute space which "in its own nature, without relation to anything external, remains always similar and immovable." [1] He certainly recognized the difficulties involved in speaking of such an unreal structure whose "parts cannot be sensually perceived" [1]; but with his basic law of dynamics

$$\text{Force} = \text{Mass} \times \text{Acceleration}$$

he was able to indicate other means to determine motion relative to space and at the same time the state of absolute space itself. If we observe the relative or "apparent" motion of an observable body, i.e., its motion relative to the observer, and measure the effective force, we can calculate the absolute motion of the body. With the aid of the laws of dynamics we can therefore determine the state of motion of the unobservable absolute space in terms of observable phenomena.

Newton developed these ideas mainly with reference to rotation. If we imagine a rotating disk in empty space, then it is kinematically impossible to determine its state of motion. This is possible, however, as soon as we take account of dynamic phenomena. An observer located on the disk could measure a centrifugal force k, i.e., the pressure that drives a body fastened to the disk toward the rim, and calculate from it the velocity of the rotation by the formula $k = m\omega^2 r$. Newton demonstrated in a very ingenious fashion how one could also determine the direction of the rotation. Carrying through his idea for our

[1] Newton, Naturalis Philosophiae Principia Mathematica, Definitions. English translation by Motte and Cajori, Berkeley, 1947, p. 6.

example, let us imagine a second smaller disk in the center of the first, rotating relative to it. On this second disk the centrifugal force will be greater or smaller depending on whether the two disks rotate in the same or in opposite directions. That direction of rotation of the second disk which causes an increase in the centrifugal force is therefore also the direction of rotation of the first disk.

Newton distinguishes consequently between the real and the apparent motion of a body. This distinction is not always obvious, and Newton regards it as the task of mechanics to develop methods that allow us to carry through the distinction between real and apparent motion in all cases. ". . . how we are to obtain the true motions from their causes, effects, and apparent differences, and the converse, shall be explained more at large in the following treatise. For to this end it was that I composed it." [1] These concluding words of the introduction to Newton's main work illustrate the contrast that may exist between the *objective significance* of a discovery and the *subjective interpretation* given to it by its author. Whereas the physical system of Newton's dynamics has become a basic part of science, which, though transformed by later developments into a higher form of knowledge, will always be retained as an approximation, his philosophical interpretation did not survive. It is only by passing through a state of absolutism in the theory of space, however, that we have been led to the deeper insights we have today. Only by disproving the Newtonian arguments were the logical conceptions developed that issued in the general idea of relativity and went beyond relativistic *kinematics* to a relativistic *dynamics*.

Even Newton's contemporary opponents, Leibniz and Huyghens, took exception to his view. They repeatedly returned to the problem of supplementing relativistic kinematics by a relativistic dynamics. Leibniz tried to reach this aim by rejecting the Newtonian gravitation as an action at a distance and attributing gravitation as well as inertia to the motion of masses relative to the surrounding ether. He would argue, for instance, that the appearance of centrifugal forces on a disk isolated in space proves its rotation relative to the ether and not relative to empty space.[2] He furthermore states expressly the

[1] *Op. cit.*, p. 12.

[2] This view is not precisely formulated by Leibniz, but it may legitimately be extrapolated from a passage in his *Dynamics* (Gerhardt-Pertz, *Leibnizens mathematische Schriften*, VI, 1860, p. 197) and also from his defense of the relativity of motion in the exchange of letters with Clarke.

aequipollentia hypothesium, the equivalence of hypotheses, for the description of a state of motion even for dynamic processes,[1] without however giving a mathematic proof. Yet his philosophical system led him to limit his dynamic relativity, for he wrote to Huyghens: "that to every object there corresponds a certain amount of motion, or, if you wish, force, in spite of the equivalence of the assumptions." He believes that for every motion there is some unique subject from which the motion originates, and comes to the conclusion "that there is more in nature than what geometry can determine, and this is not the least of the reasons which I use in order to prove that besides extension and its various aspects, which are something purely geometrical, we must recognize something higher, namely a force."[2] Therefore we cannot regard Leibniz' views as a consistent theory of the relativity of motion. He was unable to refute Newton's arguments. Neither was Huyghens able to refute them, although he found a very interesting interpretation of the centrifugal force, which however, can no longer be maintained.[3]

It was Ernst Mach who discovered the argument that can be constructed against Newton's theory of centrifugal force and which leads to a relativistic dynamics. Mach deals with the *pail experiment* described by Newton. A pail half full of water is suspended by a rope and is put in a state of rotation by twisting the rope. In the beginning the pail will not drag along the water, but will rotate alone. Gradually, however, it will drag along the water, and the surface of the water will assume the well-known hollow shape, which is due to centrifugal force. When the pail is stopped, the water continues to rotate and retains its parabolic surface. Newton concludes that the centrifugal force cannot be explained by a relative motion, since a relative motion exists between the pail and the water at the beginning as well as at the end of the process, while the centrifugal forces appear only at the end. If the pail rotates while the water is at rest, there is no centrifugal force. If, however, the water rotates while the pail is at rest, there exists a centrifugal force.

[1] Gerhardt. *op. cit.* p. 507.

[2] See Reichenbach, "Die Bewegungslehre bei Newton, Leibniz und Huyghens," Kantstudien 29, 1924, p. 432, where a detailed treatment of this problem is given. [The English translation of this paper will be included in the forthcoming volume of *Selected Essays* to be published by Routledge and Kegan Paul, London.—M.R.]

[3] Reichenbach, *op. cit.*, pp. 434f.

Mach replies that Newton overlooked the fact that the surrounding masses of the earth and of the fixed stars have to be taken into consideration. The water rotates not only relative to the pail but also relative to these large masses, which may be considered as the cause of the centrifugal force. The centrifugal force therefore does not indicate a rotation relative to absolute space, but only a rotation relative to the masses of the universe. If we consider the water with its hollowed surface as being at rest, then the earth and the fixed stars rotate around the water. In this conception the centrifugal force is a *dynamic gravitational effect of rotating masses.*. Such a force of traction originating in the rotation of masses can very well be conceived. A moving electric charge—as Einstein argued later—produces forces that do not exist when the charge is at rest.

What is new in Mach's interpretation is the idea that the inertial force can be interpreted in the relativistic conception as a dynamic gravitational effect. *Relativity can be extended to dynamics if forces are reinterpreted relativistically.* The same force that affects a body K_1 as the result of the rotation of K_1 according to one interpretation, affects it according to the other interpretation as the result of the rotation of K_2. We thus arrive at a complete reinterpretation of the concept of force (although Mach was not fully aware of it). Forces are not absolute magnitudes, but depend on the coordinate system. In physics there are both kinds of magnitudes; electric charge and entropy are *invariants*, i.e., magnitudes independent of the coordinate system; velocity and acceleration are *covariants*, i.e., depend on the coordinate system. With Mach's solution of the problem of rotation, the gravitational field is deprived of its absolute character and recognized as a *covariant magnitude* which varies with the state of motion of the coordinate system. This result, which represents the most significant aspect of Mach's view, expresses for the first time the basic idea of the principle of general covariance.

Mach draws a very interesting conclusion from his considerations. Not only the large masses of the heavenly bodies, but also smaller masses, must have dynamic gravitational effects, though to a correspondingly smaller degree. If we were to give the walls of the Newtonian pail a thickness of several miles,[1] then its rotation around the stationary water would produce a hollowing of its surface similar to that of the rotating water, but of a smaller degree. This idea was

[1] E. Mach, *op. cit.*, p. 232.

tested later experimentally by Friedländer[1] on the flywheel of a rolling mill, which should produce in the neighborhood of its axis a centrifugal field that would affect bodies not participating in the rotation of the wheel. The effect was not demonstrable, however, as it fell below the limit of accuracy. Yet Mach's claim is retained in the modern theory of relativity.

This consequence of the ideas of Mach shows, on the other hand, that the dynamic relativity of motion is more than a philosophical principle, since it leads to observable consequences. Although these consequences appear reasonable, we cannot say *a priori* whether they will be true, since this can only be decided by an experiment. We must therefore investigate Mach's solution of the problem of rotation in more detail.

Let us consider two world-systems (Fig. 36), each of which consists

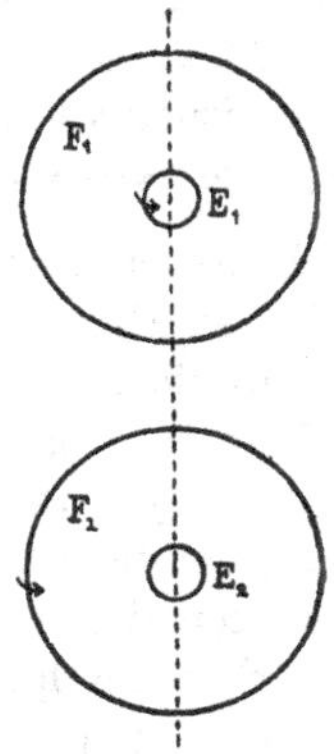

Fig. 36. Illustration of the problem of rotation.

[1] Dr. Benedict and Immanuel Friedländer, *Absolute oder relative Bewegung*, published by Leonhard Simion; Berlin 1896. These experiments were performed in 1894 by Immanuel Friedländer by means of a torsion balance. He states that he obtained the effect as expected, but was not sure whether it might not have been due to other causes. Incidentally, Friedländer is much clearer than Mach regarding the need for an empirical decision of the problem of the relativity of motion. He also recognizes "that the correct form of the law of inertia is found only if relative inertia, as a mutual effect of masses, and gravitation, which is also a mutual effect of masses, are reduced to one unified law" (page 17).

of an earth E_1 and E_2 and of a fixed star shell F_1 and F_2, respectively. The two systems are a great distance apart, but light signals can be sent back and forth to determine their relative states of motion. Let us now assume the state of motion to be as follows, using one or the other of the following two interpretations for its description (the axis of rotation is the dotted line):

Interpretation I: E_1 rotates, F_1 is at rest
E_2 is at rest, F_2 rotates

Interpretation II: E_1 is at rest, F_1 rotates
E_2 rotates, F_2 is at rest.

These two interpretations are *kinematically* equivalent. Are they also *dynamically* equivalent? Mach's theory asserts that they are, but it actually says more than that. According to the theory, it must be true that if centrifugal forces appear in E_1 then centrifugal forces must also appear in E_2, since both earths are in a state of motion relative to their respective fixed stars. This is an empirically verifiable assertion. Let us assume that it is not satisfied and that centrifugal forces appear in E_1 alone. Does this disprove dynamic relativity?

The example we have constructed corresponds to the theory of Newton, and Newton would say that it compels us to accept interpretation I and establishes absolute space. Closer examination will show, however, that this claim is exaggerated; interpretation II can also be carried through dynamically. Interpretation I states dynamically: if E rotates relative to absolute space then centrifugal forces will appear. Therefore there are centrifugal forces in E_1 and not in E_2. Interpretation II states on the other hand: if the fixed star shell F rotates relative to absolute space, it produces dynamically a gravitational field in E. There are therefore attraction forces in E_1 and not in E_2. The inferred state is the same in either case and therefore neither interpretation is dynamically incorrect. We find that *if the conditions are realized which were assumed by Newton, there exists absolute space, but its state of motion cannot be determined.*

We obtain the following result. Mach's idea (Mach's principle in the wider sense) of considering mechanical forces as covariant magnitudes, to be interpreted as inertia or dynamical gravitation according to the to the state of motion of the coordinate system, permits the application of dynamic relativity under all conditions. However, it does not exclude absolute space. This exclusion is accomplished only by a further assertion, Mach's principle in the narrower sense, connecting the appearance of centrifugal forces exclusively with the relative motion of masses, excluding therefore the state described above and requiring centrifugal forces in E_1 as well as in E_2. But this principle is undoubtedly empirical.

Mach's (and also Einstein's) relativity theory of dynamics is based therefore on the superposition of two principles, one epistemological and one empirical. The epistemological principle states that every phenomenon is to receive the same interpretation from any given moving coordinate system. The observable phenomena therefore do not single out any state of motion. The empirical principle states that all physical phenomena depend only on the relative position of bodies and not on the positions of these bodies in space. Two similar systems differently oriented in space must therefore show the same physical phenomena.

Even though attempts were made to assign a purely formal meaning to the relativity theory of dynamics according to which the relativistic interpretation describes reality, but has no claim on truth, such an untenable distinction misunderstands the significance of the Mach-Einstein theory. The relativity theory of dynamics is not a purely academic matter, for it upsets the Copernican world view. It is meaningless to speak of a difference in truth claims of the theories of Copernicus and Ptolemy; the two conceptions are equivalent descriptions. What had been considered the greatest discovery of western science compared to antiquity, is now denied its claim to truth. However much this fact may caution us in the formulation and evaluation of scientific results, it does not signify a regression in the historical development of science. The theory of relativity does not say that the conception of Ptolemy is correct; rather it contests the absolute significance of either theory. It can defend this statement only because the historical development passed through both of them, and because the conquest of the Ptolemaic cosmology by Copernicus gave rise to the new mechanics, which in turn gave us the means to recognize also the one-sidedness of the Copernican world view. The road to truth has followed in this case the purest form of the dialectic which Hegel considered essential in every historical development.

Of course, it would be an overestimation to interpret Mach's ideas as the completion of the synthesis. When Mach replied to Newton that centrifugal forces must be explained by means of relative motion alone, he did not formulate a physical theory, but only the beginning of a program for a physical theory which must eventually deal with all mechanical phenomena, not only centrifugal force. Above all it must explain relativistically the phenomena of *motion in a gravitational field*, e.g., the motion of planets. The greatest achievement of Newtonian mechanics was that it gave a dynamic foundation to the Copernican world view. While from the point of view of kinematics there existed no difference between the universes of Ptolemy and Copernicus, Newton decided in favor of Copernicus from the point of view of dynamics. It was only for this particular world description that his theory of gravitation offered a *mechanical explanation*. The complicated planetary orbits of Ptolemy, on the other hand, did not fit into any explanation. If we wish to establish the equivalence of both world conceptions, we must find a theory of gravitation sufficiently general to explain the Copernican and also the Ptolemaic planetary motion as a *gravitational phenomenon*. Herein lies the great achievement of Einstein, compared to which the ideas of Mach appear only as preliminary suggestions: Einstein has indeed found such a comprehensive theory of gravitation; and it is with this discovery, which places him on the same plane as Copernicus and Newton, that the problem of the relativity of motion has been brought to a conclusion.

§ 35. MOTION AS A PROBLEM OF A COORDINATIVE DEFINITION

Before we continue the development of the theory of gravitation, i.e., the physical part of the principle of relativity to which we are led by dynamic relativity, let us add a remark about its epistemological aspects. We are now in a position to give a rigorous formulation to the philosophical status of the epistemological assertions of the theory of relativity on the basis of previously developed concepts.

Why can motion only be characterized as relative? This question leads again to the concept of coordinative definition; the unobservability of absolute motion indicates the lack of a coordinative definition. The customary presentations have not always made this problem clear. If it is agreed that only relative motion is observable and is therefore the only objective phenomenon, this statement, which is based on the *principle of the identity of indiscernibles*, is questionable, because it assumes a metaphysical character. If we cannot recognize a difference, does it mean that there actually is no difference in the objective phenomena? This view overlooks the fact that purely logical considerations are involved. The question which system is in motion is not a defined question, and therefore no answer is possible. We are not dealing with a failure of knowledge but with a logical impossibility; the two conceptions which we are to distinguish cannot be formulated meaningfully, and consequently an answer that decides for one of them would be meaningless. To ask for the state of motion of a body is possible only if we have previously defined which system is to be the rest system; *a coordinative definition of rest* must therefore be given before we can even ask a question about a state of motion. This statement explicates the phrase "the relativity of motion." We found in § 33 that this relativity, which depends on the arbitrariness of a coordinative definition, applies even if the dynamic relativity of Mach cannot be carried through and the dynamic effects depend on the orientation of the coordinate system in space. The problem of motion therefore leads to the discovery of a new coordinative definition for space in addition to the coordinative definitions for unit length and the comparison of lengths, namely, the *coordinative definition of rest.*

It is for this reason that the idea of simplicity cannot be used to decide between the Ptolemaic and Copernican conceptions. The Copernican conception is indeed simpler, but this does not make it any "truer," since this simplicity is descriptive. The simplicity is due to the fact that one of the conceptions employs more expedient definitions. But the objective state of affairs is independent of the choice of definitions; this choice can result in a simpler description, but it cannot yield a "truer" picture of the world. That these definitions, e.g., the definition of rest according to Copernicus, lead to a simpler description, of course expresses a feature of reality and is therefore an objective statement. The choice of the simplest description is thus possible only with the advance of knowledge and can in general be carried through only within certain limits. One description may be simplest for some phenomena while a different description may be simplest for others; but no simplest description is distinguished from other descriptions with regard to truth. The concept of truth does not apply here, since we are dealing with definitions.

By basing the relativity of motion on the need for a coordinative definition, we arrive at a more fundamental relativity. The coordinative definition of rest for an entire coordinate system is generally not accomplished by mere reference to a small region. Such a definition would presuppose the concept of rigid bodies, which permits the restriction of the definition of rest to the state of motion of three points,[1] which would then determine the state of motion of the entire system. In general, however (i.e., if rigid bodies are not referred to), we must define a special state of rest for each point. We can then define an entire point system as a rest system even if its individual points change their mutual distances by comparison with rigid bodies. For example, if we define a rubber band which is stretched (in the customary sense) as being at rest, this definition will never lead to contradictions. The preference for the definition of rigidity in terms of rigid bodies is based again merely on descriptive simplicity. In order to understand this procedure, it may be remembered that the use of rigid bodies enables us to define the unit length in terms of a single space point and to extend it to other space points by the transport of rigid rods. The same procedure is applicable to the definition of rest in terms of three points in space; rest at other points in space is then defined in terms of rigid-rod connections with the original three points. This particular kind of definition is, of course, not logically necessary.

The result attained by Mach and the earlier relativists has sometimes been formulated by saying that only the relative motion of the earth and stars, not their individual motions, is an objective fact. Our present considerations force us, however, to go one step further and assert a *relativity of relative motion*. Using a crude picture, let us consider the earth as attached to the fixed stars by rubber bands. We can now define the continually stretching rubber bands as being at rest, and in this manner *transform away the relative motion* within the earth-star system. Therefore even this relative motion is not an absolute fact; it applies only to certain specific coordinate systems, namely those that can be realized by rigid bodies. If we therefore speak of an "objectively recognizable" relative motion, this can only mean a relative motion with respect to rigid bodies as defining relative rest.

[1] Even these three points cannot be given arbitrarily; they must satisfy three restricting conditions on their nine coordinates, which prescribe the mutual distances between the points. If we think of these nine space coordinates as functions of time, only six of these functions can be given arbitrarily. The assertion that a rigid body has six degrees of freedom is equivalent to this statement.

To clarify this situation, let us use an example chosen for simplicity from the two-dimensional domain. In an inertial system K of finite extension—we may think for instance of a Newtonian inertial system—let the normal space-time measurements be defined according to Einstein. Let us imagine radii r drawn from a point P_1, which cover the plane in all directions (Fig. 37) and moving points P' on every

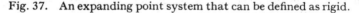

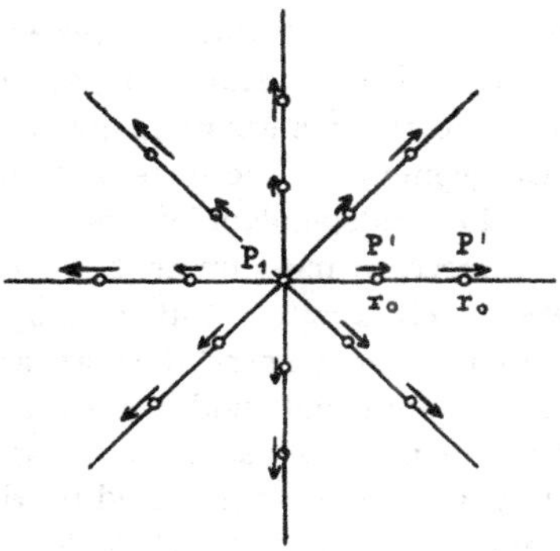

Fig. 37. An expanding point system that can be defined as rigid.

point P of this plane. Let us assume that all these points P' begin to travel along their respective radii away from P_1 at the same time $t = 0$ (defined in K in the sense of Einstein's definition of simultaneity). The velocity v of each point is uniform and $v = r_0 q$, where q is a general constant and r_0 the distance of the point P_1 at the time $t = 0$. The equation of motion of each point is therefore given by

$$r = r_0 qt + r_0 \tag{1}$$

Points lying on a circle around P_1 at the time $t = 0$ will always have the same velocity, but points lying on larger circles will move away with correspondingly higher velocities, so that the distances between the circles as well as their circumferences are continually increasing. Only the point P'_1, which lies on the point P_1, will remain at rest ($r_0 = 0$). The points P' therefore form an expanding point system when compared with a rigidly connected system. Yet we can define the system of the points P' as being internally at rest; then we have defined away the relative motion of the points P'. System K of the points P which was previously called rigid, is no longer rigid; its points move towards P_1. This system K is therefore rigid only if we use rigid bodies for our definition of rigidity.

The chosen example has another peculiarity. System P' was chosen to show quite normal properties if the light-geometry is applied to it. We can define a time measure for a clock at P_1 such that the time $\overline{P_1P'P_1}$ remains constant for light signals from P_1, although P' travels away from P_1 when viewed from K. System K' will then satisfy the light-axioms IV, 1 and IV, 2. We can also define the light-geometric measure of space so that the distance between any two points P' remains constant, although it constantly increases when viewed from K. The only difference these measurements will show compared to those in K is that the geometry of the system P' will be non-Euclidean. It is however static, i.e., independent of time, and it is impossible to detect from light-geometrical measurements that a relative motion is taking place between these points. But we may infer from the non-Euclidean character of the geometry that we are not in an inertial system and that the system is not rigid in the sense of Euclidean light-geometry, especially not in the sense of matter-geometry. We therefore have in the class of systems determined by the light-axioms I–IV a more general definition of rigidity. It is clear, however, that even these systems are special cases and that a system of differently expanding points can also be defined as being internally at rest. We are again dealing with nothing but a definition.[1]

§ 36. THE PRINCIPLE OF EQUIVALENCE

We now turn to the consequences of dynamic relativity, which goes beyond the epistemological relativity. For this purpose we must analyze Einstein's theory of gravitation, since Einstein adopted Mach's idea of dynamic relativity and developed it further. Whereas Mach restricted his investigations to rotation, Einstein applied the principle to all kinds of motions; consequently his formulation is superior. He was able to give this general formulation by transforming the ideas of Mach into a differential principle.

Einstein expressed his *principle of equivalence* in the form of a thought experiment. Let a mass m be suspended by a spring in a closed compartment such as an elevator (Fig. 38). A physicist in this

[1] For a more detailed discussion of this example see A., pp. 49 and 128. Another example of a system of radially expanding points is given on p. 60 in A., in which the light-geometry is Euclidean. In regard to the replaceability of rigid rods by the definition of rigidity of the light-geometry, see § 27 of the present book.

compartment observes suddenly that the spring expands. He can easily verify this expansion by using a measuring rod. The increase

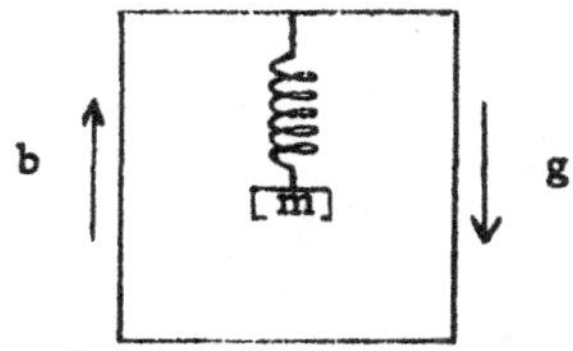

Fig. 38. Equivalence of acceleration and gravitation.

in the tension of the spring indicates a stronger pull of the mass m. How can the physicist find the cause of this pull? He could give two explanations.

Explanation I. The compartment has received an upward acceleration (in the direction of arrow b) from some external force. The effect of the inertia of the mass m is therefore a downward pull opposite to the direction of the acceleration.

Explanation II. The compartment has remained at rest, but a downward directed gravitational field g (arrow g) has arisen and therefore exerts a stronger pull on the mass m.

It is impossible to decide experimentally between these two explanations inside the compartment. This is still true if we permit the physicist to look out of a window, since he will observe only kinematic phenomena, and these do not enable him to decide between the two explanations. It might be objected that explanation II requires the appearance of large observable masses below the compartment, but this is true only if static gravitational fields are assumed. As soon as we admit dynamic fields in Mach's sense, the gravitational field g can be attributed to a motion of the surrounding masses.

What is the basis of this indistinguishability? According to Einstein, its empirical basis is the equality of gravitational and inertial mass. This new distinction must be added to the usual distinction between mass and weight. There are therefore three concepts: inertial mass, gravitational mass, and weight.

The first distinction originated with Newton's discovery that the weight of a body depends not only on the body itself but also on the distance at which the body is located relative to the attracting mass. A mass m (Fig. 39) resting on a spring balance will exert a different force (measurable by the tension of the spring, which is indicated by

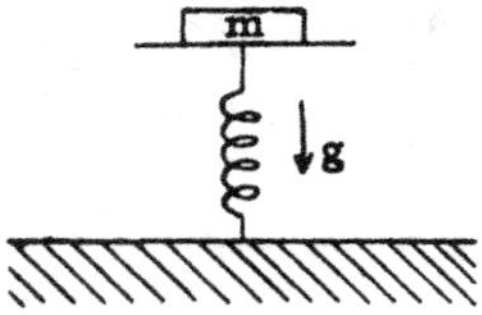

Fig. 39. Measurement of gravitational mass.

its length) on the support, according to the distance of the apparatus from the center of the earth. This fact is expressed by the formula

$$\vec{F} = m.\vec{g} \tag{1}$$

which resolves the force $\vec{F}$ exerted by a body on its support into the intensity $\vec{g}$ of the earth's gravitational field (the vectorial nature of $\vec{g}$ is usually ignored and it is written "g") and a proportionality factor m due to the body itself. The structure of formula (1) is analogous to that of the formula

$$\vec{F} = e.\vec{E} \tag{2}$$

of electrostatics, where the mechanical force $\vec{F}$ results on the one hand from an intensity $\vec{E}$, which is independent of the attracted body and characterizes the field, and on the other hand a proportionality factor e which is interpreted as the electric charge of the body. Correspondingly we might call m the *gravitational charge.*[1] This factor m is the *gravitational mass* of the body, i.e., the constant that expresses the effect of gravity upon it.

The mass of the body has also a quite different effect. If a carriage supporting the mass m is put in motion on a horizontal plane by the release of a compressed spring (Fig. 40), then the force $\vec{F}$ of the spring will produce a certain acceleration $\vec{b}$ which determines the velocity with which the carriage continues to roll horizontally after the push. The following equation applies to this relation

$$\vec{F} = m.\vec{b} \tag{3}$$

[1] H. Weyl, *op. cit.*, p. 225.

It turns out that in this equation m has the same value as that in equation (1). This is an empirical statement which we can imagine to be tested as follows. Assume objects of different materials, which, according to Fig. 39, show the same compression of the spring and are then pushed, according to Fig. 40, by a spring under the same

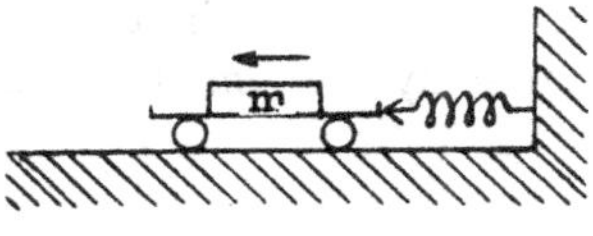

Fig. 40. Measurement of inertial mass.

tension. It can be shown that the push will give them equal velocities. This result is not self-evident. It is conceivable, for example, that volume would have an influence on inertia and that among the masses of equal weights those having a greater volume would receive a smaller velocity in the experiment of Fig. 40. This question can be decided only by experience.

The principle of the equality of inertial and gravitational mass, which incidentally is also the reason for the equality of the velocities of falling bodies (a body which is more strongly attracted by gravity has to overcome a correspondingly greater inertia) has been confirmed to a high degree by experiments. It is mentioned explicitly by Einstein as an empirical principle constituting the basis of his principle of equivalence.

The equivalence of inertia and gravity is the strict formulation of Mach's principle in the narrower sense. It implies that every phenomenon of inertia observable in an accelerated system can also be explained as a gravitational phenomenon; therefore it cannot be interpreted to indicate uniquely a state of motion. Conversely, we can use the principle of equivalence to *transform away* that gravitational field which was considered an absolute datum in classical mechanics. A freely falling elevator is a system in which the gravitation of the earth is transformed away. Any object in it when pushed would assume a rectilinear, force-free motion in the sense of the law of inertia.

The possibility of "transforming away" is subject to certain essential restrictions. Generally speaking, we can transform away gravitational fields only in infinitesimal regions. Let us consider for example the radial field of the earth (Fig. 41). If we let a rigid system of cells (the dotted lines of the figure) move in the direction of arrow *b* with an acceleration $g = 981$ cm/sec², the earth field will be transformed away in cell *a* but not in any of the others. We can now make the following statement: for any given small region *b* we can always specify for the system of cells an accelerated motion which will transform away the gravitational field at *b*. We may therefore say that any gravitational field can always be transformed away in any given region, but not in all regions at the same time by the same transformation.

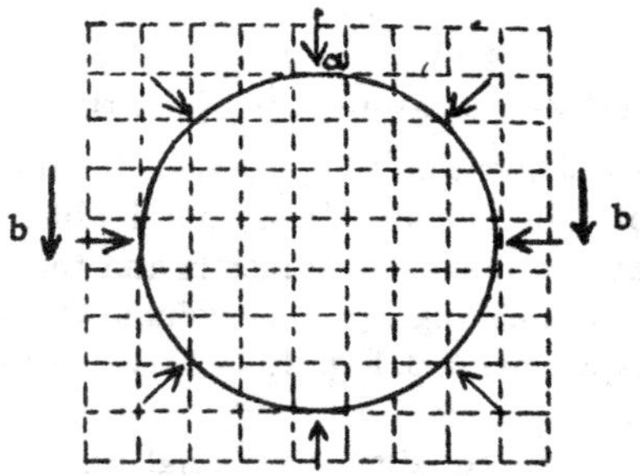

Fig. 41. Local "transforming away" of the gravitational field.

This principle takes the place of the Newtonian concept of inertial system. By inertial system [1] Newton understands those astronomically determined systems in which the law of inertia applies, i.e., those systems that move uniformly relative to absolute space. It can be shown within the framework of Newton's theory that one can obtain local inertial systems by transforming away the gravitational field, although these systems are in a different state of motion provided that the equivalence of inertial and gravitational mass is presupposed. The gravitational field, which as such is still present, is compensated in these local systems by their acceleration relative to absolute space and the resulting inertial forces. According to Einstein, however, only these local systems are the actual inertial systems. In them the field, which generally consists of a gravitational and an inertial component, is transformed in such manner that the gravitational component

[1] This term was introduced by L. Lange, "Über die wissenschaftliche Fassung des Galileischen Beharrungsgesetzes", *Wundts Philos. Studien*, 1885, Vol. II.

disappears and only the inertial component remains. There are, strictly speaking, only local inertial systems. The astronomical inertial systems of Newton can at best be approximations which gradually change in the neighborhood of stars. Only because distances in space are large compared to the masses of the stars, and because the stars have very low speeds, are astronomical inertial systems possible as approximations.

We must now formulate this idea more precisely. Above all we have to state exactly what is meant by an "actual" inertial system, which for the time being has only a more or less intuitive meaning. Let us investigate first how the local inertial systems result according to Newton. Newton's equation for the motion of a mass point in a gravitational field is given by

$$\ddot{x} = g \tag{1}$$

If we now relate the x-coordinate to a freely falling system, i.e., introduce the transformation

$$\begin{aligned} x &= x' + \frac{g}{2}t^2 \\ y &= y' \end{aligned} \tag{2}$$

then

$$\ddot{x} = \ddot{x}' + g$$

and (1) becomes

$$\ddot{x}' = 0 \tag{3}$$

which is the equation of motion in an inertial system. Within mechanics there exists no difference between the two kinds of inertial systems, and it would be a play on words to argue that one or the other of the two is an "actual" inertial system. If we take into account, however, *extra-mechanical* phenomena, there will be a difference: *whereas according to Newton the astronomical inertial systems form the normal systems for all phenomena, Einstein maintains that it is the local inertial systems which form the normal systems.* We shall study the resulting difference in the example of the motion of light.

According to the Newtonian theory only the astronomical inertial systems are the normal systems for the propagation of light. Only in them does light travel in straight lines, while its path is curved in a local inertial system. The motion of a light ray which moves parallel to the y-axis is given in the Newtonian inertial system by the differential equation

$$\begin{aligned} \dot{x} &= 0 \\ \dot{y} &= c \end{aligned} \tag{4}$$

These equations are valid according to Newton even if there is a gravitational field, as for example on the surface of the earth. The earth is embedded (for short intervals of time) in an astronomical inertial system upon which the gravitational field of the earth is only locally superimposed. With respect to light, this gravitational field does not exist at all. If we now apply transformation (2) to these equations, they become

$$\begin{aligned} \dot{x}' &= -gt \\ \dot{y}' &= c \end{aligned} \tag{5}$$

Relative to K', light no longer travels along straight lines, because its x'-coordinate is no longer a linear function of time.

According to Einstein, however, the local inertial systems are the actual inertial systems for all other phenomena. In the case of the light ray, for instance, the equation of motion must be linear in the local inertial system K', and the differential equations must therefore be:

$$\begin{aligned} \dot{x}' &= 0 \\ \dot{y}' &= c \end{aligned} \tag{6}$$

If we now go in turn with transformation (2) to the system K which is stationary on the earth's surface and consequently at rest in the astronomical inertial system, the equations will become

$$\begin{aligned} \dot{x} &= gt \\ \dot{y} &= c \end{aligned} \tag{7}$$

It is relative to *this* system that light is now curved.

We shall illustrate the train of thought that leads from (6) to (7) by the path of a light ray; this will bring out the purely kinematic basis of the inference. Let us imagine a compartment (Fig. 42) at

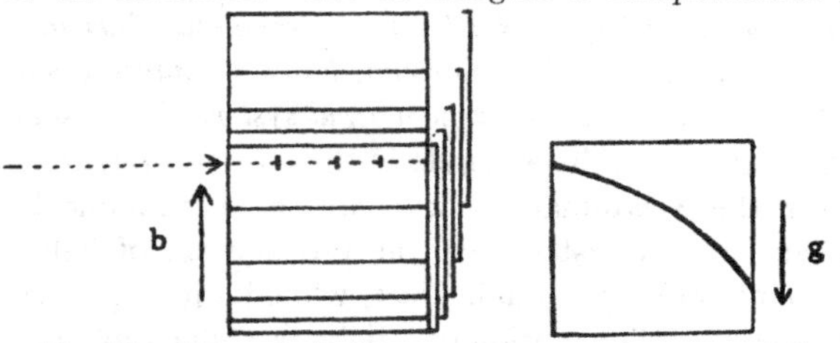

Fig. 42. Bending of a light ray as a consequence of the principle of equivalence.

rest on the earth. Relative to the local inertial system it will perform an upward accelerated motion. Let us also assume that a light ray enters the compartment through a slit on the left-hand side. We can now determine its path within the compartment if we assume that the local inertial system is at rest, and if we construct the motion of the light ray relative to the compartment by superimposing the straight line path of the light ray upon the accelerated motion of the compartment. The different consecutive positions assumed by the compartment are indicated by the square brackets of Fig. 42. The end of the light ray is a little farther to the right for each successive position of the compartment, corresponding to the marks on the dotted line. It can now easily be seen that these marks have different positions relative to the compartment in its various locations. On the right-hand side we have drawn the same process relative to the compartment as a rest system and indicated the marks this time in their relative positions in the compartment. The path of the light ray is therefore a *curved line* relative to the compartment. This is a purely kinematic effect. It derives from the fact that the horizontal motion of the light is uniform, while the vertical motion of the compartment is accelerated. Since we have started from the assumption, however, that light travels in straight lines relative to the local inertial system which falls freely relative to the earth, we have now arrived at the far-reaching physical consequence that light assumes a curved path relative to a system which rests on the earth: there is a curvature of light in the gravitational field of a mass center.

It is irrelevant in this case whether the mass center itself is resting in an astronomical inertial system, since this inertial system no longer constitutes a normal system in the neighborhood of the mass center. Indeed, it is no longer reasonable to speak here of an inertial system with a superimposed gravitational field. The astronomical inertial system is destroyed in the neighborhood of the mass center and cannot be extended from the surrounding space to the region of the mass field without losing its inertial character. Its functions have been taken over by the local inertial system to which it cannot be rigidly attached.

In these assumptions we find the core of the general theory of relativity. It is a genuine physical principle which, with the inclusion of all nonmechanical phenomena in the characterization of the local inertial system, states a *physical hypothesis* that goes far beyond the experience stated in the equivalence of inertial and gravitational mass. Einstein's hypothesis corresponds to a methodological procedure frequently used in physics. Although the hypothesis does not follow logically from the empirical evidence but claims much more, it is assumed in the hope that the observation of further derivable consequences will confirm it. After the special theory of relativity had formulated the laws of clocks, measuring rods, the motion of light, etc., for inertial systems, the new hypothesis could now be formulated by the statement that it is not the astronomical inertial systems, but the local inertial systems, for which the special theory of relativity holds. The gravitation-free ideal case required for the special theory of relativity is therefore not realized in the astronomical inertial systems, but in the local inertial systems. We may thus speak of the *principle of local inertial systems*, which states that the *local inertial systems are those systems in which the light- and matter-axioms are satisfied.*[1] With this hypothesis Einstein introduces the general theory of relativity, and the special theory of relativity thus becomes the limiting case of the general theory.

For the sake of completeness, we shall now show how the same inferences that lead to physical consequences regarding light also lead to similar consequences regarding clocks. We shall again consider a kinematic effect that results from the accelerated motion of a clock relative to an inertial system, and infer from it an effect in the gravitational field. The kinematic effect with which we are concerned is the *Doppler effect.*

Let us first consider the Doppler effect that results from uniform motion (Fig. 43). Let us assume that an observer is moving in a

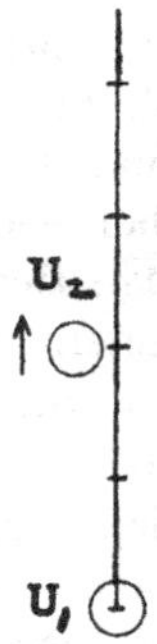

Fig. 43. Doppler effect as a result of uniform motion.

[1] Strictly speaking this should read: "in which these axioms are satisfied to a higher degree of approximation". Cf. A., § 34.

straight line with uniform velocity away from U_1. Whenever the clock U_1 completes a period, it sends out a signal which will reach the observer at increasingly distant points. The intervals between the various light signals are therefore longer for the observer than the unit intervals of the clock U_2 which he carries with him. For him clock U_1 runs slower than U_2. Let us now consider a similar process in the case of accelerated motion (Fig. 44). The two clocks U_1 and U_2 are

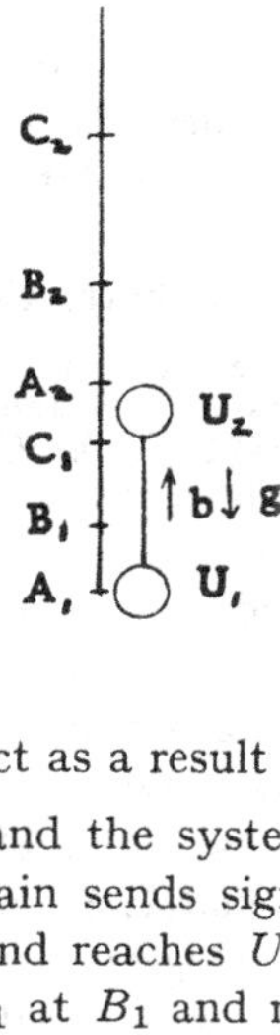

Fig. 44. Doppler effect as a result of accelerated motion.

connected by a rigid rod, and the system which they form has an accelerated motion. U_1 again sends signals after each unit period. The first signal leaves A_1 and reaches U_2 when U_2 has reached A_2. The second signal leaves U_1 at B_1 and reaches U_2 at B_2, etc. The distances A_1A_2, B_1B_2, C_1C_2 . . . will become longer and longer, and an observer who moves with U_2 will thus experience a Doppler effect in the sense of a retardation of U_1. In either case there is therefore a retardation of one clock relative to the signals which arrive from the other clock. Whereas in the case of uniform motion, only one of the clocks is in motion while the other is at rest, the effect will appear in the case of accelerated motion even when the two clocks are at rest relative to each other, provided the rigid system which they form moves as a whole. The latter case permits reinterpretation in terms of the principle of equivalence. Two clocks which are at rest in the gravitational field of a mass center are in an accelerated motion relative to the corresponding local inertial system. Our consideration will therefore lead directly to the assertion that a gravitational field produces a retardation of those clocks which are located in regions that have a higher absolute value of the gravitational potential. In the case of atom clocks, there would be a red shift of the spectral lines, because a retardation of the frequency manifests itself as a shift of the wave-length in the direction of the red end of the spectrum.

It should be noted that this effect is independent of the retardation of clocks discussed in § 30. We have used for its derivation nothing but the Doppler effect. The Doppler effect was also known in the classical theory of time, which does not include however the retardation of clocks discussed in § 30. The retardation of clocks in a gravitational field must therefore occur if the principle of equivalence alone is correct, regardless whether there is an Einsteinian retardation of clocks for uniform motion. This latter effect shows only in the *quantitative* calculations of the retardation of clocks in a gravitational field, where it appears as a small correction factor.

This last result is due to the fact that the Doppler effect can be calculated as the superposition of two effects, namely, the classical Doppler effect and the Einsteinian retardation of clocks. Conversely, we can recognize from this result that the Einsteinian retardation of clocks in uniform motion has nothing to do with the Doppler effect.

The bending of light and the retardation of clocks are direct consequences of the principle of equivalence, and they demonstrate very clearly the hypothetical character of the principle since they are empirically confirmable phenomena. The third of the so-called Einstein effects, namely the advance of the perihelion of planetary orbits, does not follow immediately from the principle of equivalence, but from Einstein's theory of gravitation based upon it, especially from the field equations to be mentioned in § 39.

§ 37. EINSTEIN'S CONCEPT OF GRAVITATION

Continuing the investigation of the principle of the local inertial systems, we are led to a concept of gravitation that is much more complicated than that of Newton.

According to Newton the gravitational force is given by the expression $\frac{m_1 m_2}{r^2}$; this formula resulted from the fact that he considered the gravitational field as superimposed on an inertial system, a conception that enabled him to measure and describe the field in terms of the coordinates of the inertial system. The given expression for the gravitational force holds only if the system of reference is an inertial system. In Einstein's conception, however, the gravitational field cannot be measured relative to an inertial system, since the gravitational field is no longer considered as a phenomenon superimposed upon the inertial system, but as a region in which there exists no extension of the astronomical inertial systems. If local inertial systems are sought within such a region, gravitation must be transformed away locally, and consequently it is impossible to find an inertial system relative to which a gravitational field could exist and be measured. Einstein's gravitational field must therefore be formulated without reference to a unique coordinate system.

This assumption agrees with the ideas of Mach, who looked upon gravitation as a covariant magnitude, the expression of which transforms with the coordinate system. In every system of reference deviating from the local inertial systems, a formulation of the gravitational field must be possible. Furthermore, no system is distinguished from the others as the one relative to which we could measure a "true" gravitational force. We must therefore look for a mathematical expression for gravitation sufficiently "elastic" to achieve such a general characterization.

A scalar theory of gravitation can no longer accomplish this task. Such a theory characterizes the gravitational state at every point by a single number, the *potential*; and the gravitational force will then be characterized by the *potential gradient* which can be calculated for every point from the potential field; thus no further parameters are required.

The new theory has to accomplish considerably more. Let us take a less simple system of reference, e.g., a rotating disk. All mechanical phenomena observable on the disk must be interpretable, according to our principle, as gravitational effects. The centrifugal force, which increases with the first power of the distance from the center, might still be represented by a law of potential, although it would not correspond uniformly to the fundamental differential equation of a potential field, $\Delta\phi = 0$, since the center constitutes an exception to this condition. This force, however, is not the only one effective on the disk. An observer located on the disk would also notice the effects of the so-called Coriolis force, which exerts a lateral pull on moving objects, for instance, the deviation to the right of a projectile under the influence of the rotation of the earth. The new mathematical expression for gravitation must be comprehensive enough to account for both the centrifugal and the Coriolis force, and this task cannot be accomplished by a scalar potential.

For this reason we have to use a *tensor potential*, which means that the gravitational state of a point is no longer characterized by a single parameter, but by a set of parameters, a system of tensor components, which accounts for all forces occurring. The simple case of the Newtonian field, in which one parameter is sufficient, is then contained in the new conception as a special case. It is reasonable to assume that the Newtonian field can occur only as an approximation and will never be strictly realized. That astronomers have managed so far with a single parameter for each point (e.g., in the earth's field) is due to the fact that one of the tensor components predominates over the others and hence has been the only one noticed so far. The new expression therefore formulates not only the covariance of gravitation but also leads to a qualitative and quantitative change of the gravitational field even in "normal" cases, which should be noticeable in very precise measurements. This consideration shows that we are dealing in Einstein's theory of gravitation with a physical theory which is suggested by epistemological considerations but not derivable from them.

According to this theory, the gravitational state at a point is best compared with the state of tension existing, for example, in a beam of a bridge. In every supporting member of a bridge there is a state of tension producing stresses in the individual volume elements. This state cannot be characterized simply by the coordination of a force, i.e., a vector, to every point, but requires a more complicated formulation. If we assume that the beam is cut along a slanted plane, the two halves will shift relative to each other, i.e., obey a force. The magnitude and direction of this motion will depend on the direction of the plane along which the beam has been cut; each cut, therefore, determines a specific force. For this reason we cannot determine the state of the internal stresses at a point by coordinating to it a single vector. The situation is much more complicated. For each point there are infinitely many such coordinated vectors, one for each plane element that passes through the point. This situation can be simplified by means of the concept of tensor component. We do not require infinitely many vectors: it is sufficient to specify three surface elements and their corresponding vectors. Then we can calculate the vector corresponding to any other plane through this point according to the laws of the transformation of components. Since each of the fundamental three vectors is given by three components (in a three-dimensional space), the state of the internal stresses of a beam requires $3 \times 3 = 9$ specifications for each point. Furthermore, it is shown in mechanics that these nine tensor components must satisfy conditions of symmetry which reduce the number of independent specifications to 6.

The gravitational field must be conceived in a similar fashion. The characterization is complicated by the fact that it must also account for the state of motion of mass points. A mass point carried by a gravitational field does not remain at rest, like the volume elements of the beam, but falls "down." The representation must also include changes in time, and the gravitational field, therefore, is characterized by a space-time tensor, not by a space tensor. Such a space-time tensor has $4 \times 4 = 16$ and not $3 \times 3 = 9$ components. These can be reduced, as before, to 10 independent numerical terms, because of conditions of symmetry. To make the analogy with the beam complete, we have to use the Minkowskian four-dimensional representation, in which the state of a mass point at rest is represented by a *vertical straight line*, the world-line of the point. Under the influence of a tensor field this straight line will bend just like the beam of a bridge, but such a *curved* world-line is that of a point *in motion*. For this reason the motion of a point can be characterized by a tensor field.

The concept of weight will also be subject to certain changes. In Newtonian mechanics, weight results from the single gravitational force which pulls the body down at all points. In Einstein's mechanics, on the other hand, the body is in a "state of stress" due to the gravitational field; it is subject to tension and compression in all directions. These may now be combined in a resultant which we call the weight. Newtonian mechanics knows only this resultant.

Conversely, the given presentation makes it clear why, through the four-dimensional tensor characterization, the gravitational field becomes a coordinate magnitude, i.e., a quantity varying with the coordinates. We have characterized the tensor by its 4×4 components and not by the infinite number of vectors of which it is composed. If we change the coordinate system, we also change to a different set of fundamental components of the tensor, i.e., among the infinite number of vectors we choose a different set of fundamental components.

In three dimensions this would mean that the three basic surface elements have been chosen differently. In the four-dimensional case it includes also the state of motion, since the slanted direction of a rising world-line indicates motion. The introduction of a coordinate system in a different state of motion signifies therefore merely the resolution of the tensor into a different set of components.

This consideration leads to a distinction which we have touched upon several times before and which expresses a basic idea of modern science. The system of the tensor components is *covariant*, i.e., it has a different numerical composition for each coordinate system. Yet we express in this fashion a state that is independent of the coordinate system, i.e., an *invariant* state. The tensor as a whole is an *invariant magnitude*. We can recognize this property from its representation by means of components, since the components can be calculated for *every* coordinate system, if they are known for *one*. It is unfortunate that the physical terminology does not reflect this well-defined mathematical distinction. By "gravitational field" we understand the system of components of the tensor in each case; this makes the gravitational field a *covariant* magnitude. No particular term has been accepted for the invariant tensor field as a whole. It might best be called the *metrical field*, in accordance with some ideas which we shall discuss later; in fact, this term has occasionally been used with this meaning. In this terminology the gravitational field is the particular system of components into which the metrical field has been resolved.

This representation explains why the gravitational field can be transformed away. For this purpose, one resolves the metrical field in such a manner that the components, the gravitational potentials, become independent of the coordinates, i.e., become constants (this is always possible at least for local regions); then there exists no gravitational gradient. The disappearance of the gradient is then called "the disappearance of the gravitational field." There are actually three concepts involved in this problem: the tensor as a whole or the metrical field, the particular set of tensor components or the gravitational field in the wider sense, and finally the particular set of gradient coefficients of the tensor components or the gravitational field in the narrower sense. The latter two are distinguished from each other as the concepts of potential and of gradient and can therefore be distinguished as *gravitational potential field* and *gravitational gradient field*. Only the gradient field can be transformed to zero, in which case the potential field becomes constant.

In the mathematical representation, the metrical field is given by the tensor g, the gravitational potential field by the particular set of components $g_{\mu\nu}$, and the gravitational gradient field through the Riemann-Christoffel symbols $\Gamma^{\tau}_{\mu\nu}$, which are obtained from the $\frac{\partial g_{\mu\nu}}{\partial x_\tau}$. The $\Gamma^{\tau}_{\mu\nu}$ do not form a real tensor, only a linear tensor, and can therefore all at once be transformed to zero by nonlinear transformations.

A fourth concept has occasionally been introduced. We set $g_{\mu\nu} = \overline{g_{\mu\nu}} + \gamma_{\mu\nu}$, where $\overline{g_{\mu\nu}}$ are the normal orthogonal values of the $g_{\mu\nu}$, and we refer to the $\overline{g_{\mu\nu}}$ as the *inertial field* and only to the $\gamma_{\mu\nu}$ as the gravitational potential field. The $\Gamma^{\tau}_{\mu\nu}$ may then be considered as the derivatives of the $\gamma_{\mu\nu}$, since the $\overline{g_{\mu\nu}}$ as constants do not contribute to the gradient field. This resolution into inertial and gravitational field is an adaptation to the terminology of Newtonian mechanics, however, and is therefore hardly appropriate.

Finally, we can give an exhaustive formulation of the ideas of dynamic relativity using the concepts developed above.

For any given coordinate system there exists a certain gravitational field. For any coordinate system in a different state of motion, there exists a different gravitational field. If we wish to carry through a general relativity of all coordinate systems, i.e., a general application of relativity to kinematically equivalent descriptions of states of motion, the gravitational field must be correspondingly specified for every coordinate system. The coordinate systems themselves are not equivalent, but every coordinate system with *its* corresponding gravitational field is equivalent to any other coordinate system together with *its* corresponding gravitational field. Each of these covariant descriptions is an admissible representation of the *invariant* state of the world.

§ 38. THE PROBLEM OF ROTATION ACCORDING TO EINSTEIN

We shall now investigate the problem of rotation, on the basis of Einstein's idea that a special gravitational field has to be assumed for every coordinate system.

If we consider, on the one hand, the earth as rotating, this motion must be relative to one of the Newtonian inertial systems which can be introduced as an approximation. Relative to this coordinate system there exists no gravitational field, only an inertial field. If we consider, on the other hand, the earth to be at rest, there must be a tensorial gravitational field relative to a system of axes to which the earth is rigidly connected. This gravitational field manifests itself as a rotational field, which might be compared with the rotational field of a three-phase electric current whose three phases run through a correspondingly divided coil. Just as such a field whirls iron filings in its interior, so the tensorial gravitational field moves the fixed stars around in a circle. The stars will thus move with the same velocities around the axis of the field. At the same time, light rays, which are deflected in the gravitational field like heavy bodies, are rotated. Only the introduction of such a rotational field leads to the dynamic equivalent of the rotating coordinate system. A number of objections [1] that have been raised against this conception will be discussed in the following paragraphs.

1. The equivalence of rotating coordinate systems introduces velocities above the velocity of light. Points of the coordinate system that lie farther out will have increasingly higher peripheral velocities and points which lie outside of a circle with radius $r = \frac{c}{\omega}$ (ω is the angular velocity) will therefore have a peripheral velocity ωr greater than c relative to the coordinate system. The planet Neptune will already have a velocity greater than that of light if we consider the earth at rest; the fixed stars will have even greater velocities. This consequence appears to contradict a requirement of the theory of relativity.

The conclusion is erroneous, however, since we are here dealing with a problem in the general and not in the special theory of relativity. In the special theory we can maintain the velocity $3 \cdot 10^{10}$ cm/sec. as the limiting value because we allow only certain coordinate systems relative to which all velocities are to be measured. If we admit arbitrary coordinate systems, the number $3 \cdot 10^{10}$ cm/sec. can be exceeded. The limiting character of the velocity of light, however, can be maintained even in the general theory of relativity, yet the assertion is formulated differently. Given any two mass points, light signals will be the fastest connection between them. *Light is the fastest messenger*; it moves faster than any other means of communication from the same place and at the same time. We formulated this property above with the aid of the concept *first-signal*. This principle is satisfied even if we assume that the coordinate system is rigidly connected to the earth. A light signal sent from the planet Neptune

[1] See also the discussion between Wulf-Reichenbach-Anderson in *Astron. Nachr.* Vol. 213, Nos. 5083–84, 5107, 1921; Vol. 214, No. 5114; Vol. 215, No. 5154, 1922.

along a tangent to the planet's orbit moves faster than the planet itself and will run away from it. Here light has an even greater velocity, although the planet itself already exceeds the velocity $3 \cdot 10^{10}$ cm/sec. This value has therefore no significance if completely arbitrary space-time measurements are admitted.

In addition there results another restriction from the limiting character of the velocity of light. The system of axes rigidly connected to the earth can ideally be extended indefinitely, but outside the circle $r = \frac{c}{\omega}$ it is impossible to realize the axes materially. Outside this circle there can be no material points which are at rest relative to the coordinate system; in this case, the coordinate system is no longer a *real system*.[1] Neptune has its high peripheral velocity therefore only relative to ideal rest points, but not relative to points which can be realized materially. This restriction is due to the fact that in the immediate neighborhood of Neptune, as at every point, the special theory of relativity holds in infinitesimal regions. This requirement leads to an important distinction. Not all kinematically possible systems can be realized by material structures at rest relative to them. We must therefore distinguish between *real and unreal systems* (see § 41).

2. According to the general relativity of rotation, we can consider not only the earth but also any given rotating system, e.g., a merry-go-round, as the rest system. This conception, however, has absurd consequences. The horse, which in the usual interpretation pulls the merry-go-round, must in the second interpretation be able to put the earth, even the universe, in motion by means of treading, since now the merry-go-round remains at rest. How can the horse have the strength to do so?

This objection overlooks the fact that, in the relativistic conception, the rotation of the stars is due to a gravitational rotational field, and not to the horse. The latter has an entirely different task: it prevents the merry-go-round from following the rotational field and taking part in the general rotation. We see that even according to the relativistic interpretation, the horse has to perform a task determined by the mass of the merry-go-round and not by the mass of the stars. If an elevator glides down slowly and a fly inside crawls upward so that it remains at the same level relative to the building, it has to transport only its own mass—it does not have to "push down" the elevator.

[1] See § 45 of this book and A., § 46.

3. Even if the horse of the merry-go-round does not need the strength to rotate the stars, it is still the cause of this rotation, since the rotation of the stars relative to the merry-go-round begins only when the horse begins to run. The starting of the horse must therefore be the cause of the rotational tensor field. A long time, many years, must then pass before the effect of the field will reach the fixed stars, due to the limiting velocity of causal propagation. But we know that this is not the case.

This objection is not answered by the remark that velocities above that of light are possible in gravitational fields. According to our previous discussion, any effect originating on the earth cannot reach Sirius faster than a light signal, and light would take a good deal longer than our example permits. The mistake must therefore arise in a different manner.

The error is due to the fact that the causal question has been posed incorrectly. The objection starts from the assumption that we must find a cause for every change that occurs relative to the coordinate system. This assumption leads to the consequence that for every change of the coordinate system there must be a change in the causal chains, and that we must look, in our example, for a causal chain from the horse to the fixed stars. On this requirement the causal chain, like the gravitational field, would be a covariant concept. This consequence, however, constitutes the mistake: *the cause-effect relation is not a covariant but an invariant concept.* Changes of the coordinate system will not affect the status of the causal chains; they are invariant sequences which will always connect the same identical point events, independent of the kind of description. In our example, the causal sequence goes from the horse to the merry-go-round and not to the fixed stars; and that remains true if we use a coordinate system for which the merry-go-round is at rest.

If we now look for the cause of the gravitational rotational field, we must look for the dynamic equivalent of this field in a kinematically different conception, namely the usual one. In this conception the fixed stars are at rest and determine an inertial field which pervades the entire space and lends inertia to every moving mass point. The same causal connection applies to the relativized conception: the inertial field corresponds to the gravitational rotational field, which again must be caused by the fixed stars. Through their motion the rotating fixed stars produce a rotational field that in turn perpetuates their rotation. The same relations that represent the inertial field as an effect of the masses of the stars, represent in the relativized conception the rotational field as an effect of the masses of the stars. That we speak sometimes of an inertial field, at other times of a rotational field, is only a difference in *description*, not a difference in the facts described. We must therefore not look for a difference in these facts, if we go from one description to another. It is meaningless, therefore, to look for a cause of the origin of the rotational field at the moment when the merry-go-round begins to move. If we continue to relate all measurements of space to the merry-go-round, we must describe the rest of the world in a different language. We are compelled to do so because of a cause affecting the merry-go-round and not the rest of the world. With the choice of a different language we do not imply that anything has changed in the rest of the world, but only that a change has taken place between the world and our system of reference.

4. A further objection maintains that the rotational field is an "imaginary" or "fictitious" force field; we cannot demonstrate its real existence, and it is not real like ordinary gravitational fields. This objection is based on a misinterpretation of covariant concepts. The state of affairs described by the covariant concept of gravitation can be represented differently in different coordinate systems. Each of these descriptions presents the objective state in a particular way. The totality of these descriptions, however, defines an invariant situation, whereas *one* description gives only one component of the situation, so to speak, namely, its projection on a particular coordinate system. Among these components there is no difference with regard to truth. Just as we can demonstrate the ordinary gravitational field of the earth by means of the pressure which a body exerts on its support, we can also demonstrate the rotational field objectively by means of the centrifugal and Coriolis forces.

§ 39. THE ANALYTIC TREATMENT OF RIEMANNIAN SPACES

Before we continue our discussion of the problem of gravitation, we will insert a section carrying the mathematical treatment of general geometry a little further. Familiarity with these mathematical methods is indispensable for an epistemological study of space. It will be seen that the mathematical development is not as difficult to understand as is commonly believed by the nonmathematician.

We shall use here the analytic treatment of the problem of space that was introduced by Riemann and Gauss (§ 2), developing briefly the basic ideas of their mathematical methods. Let us imagine a plane subdivided by a network of curves which are numbered. Then we can indicate the position of every point of the plane by means of the small plane figure in which it lies. Point P of Fig. 45, for example, lies in

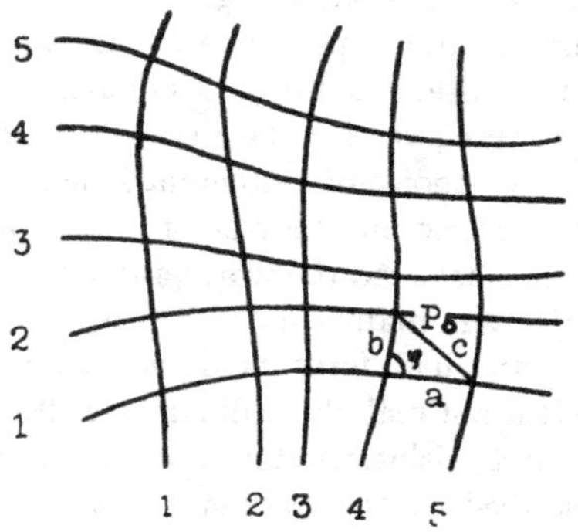

Fig. 45. Specification of a metric for curvilinear coordinates.

the plane figure 4—5/1—2. If we wish to give its position more exactly, we merely have to subdivide the mesh of the network further; thus we might estimate its position as 4.7/1.9. The curved lines form a coordinate system since they supply a numbering of all points in the plane. And this is all we can demand of a coordinate system.

We ask: how long is the diagonal c of the coordinate cell under consideration? If we had used rectilinear orthogonal coordinates, as on ordinary graph paper, instead of curvilinear coordinates, this question could be answered easily. Since the coordinates of the endpoints are (5, 1) and (4, 2), their distance is given by

$$\sqrt{(5-4)^2+(2-1)^2}$$

according to the Pythagorean theorem. This answer, however, does not apply to curvilinear coordinates, since these coordinates are nothing but identification numbers, like the numbers on houses, and do not refer to a measure of length. How can we avoid this difficulty?

Let us call the actual lengths of the sides of the coordinate cell a and b; these two sides together with the diagonal c form a skew triangle.

Such a description considers a and b as straight lines; this approximation becomes more accurate the finer the network. The following calculations should therefore be understood to be correct for the limiting case $a = b = 0$. According to the "extended theorem of Pythagoras" we may now write

$$c^2 = a^2+b^2-2ab \cos \phi \tag{1}$$

If we now let the variable x_1 range over the numbers going to the right and the variable x_2 over the numbers going upwards, then a and b will be in some relation to the coordinate differences $dx_1 = (5-4)$ and $dx_2 = (2-1)$. They are of course not identical with these coordinate differences, since these differences are derived only from identification numbers, not from a measurement of distances. The length of the segment a is not equal to 1 as is the coordinate difference dx_1; it is smaller. We must therefore write

$$a = \alpha dx_1 \qquad b = \beta dx_2 \tag{2}$$

where α and β are factors. Substituting according to (2) in (1), we obtain

$$c^2 = \alpha^2 dx_1{}^2+\beta^2 dx_2{}^2-2\alpha\beta\, dx_1\, dx_2 \cos \phi \tag{3}$$

in which α, β and $\cos \phi$ are numbers that are characteristic for the cell containing P. In different cells of the network these numbers would have different values, while the expression (3) would remain the same. We must therefore consider these numbers as functions of the position. Using the abbreviation

$$\alpha^2 = g_{11} \quad \beta^2 = g_{22} \quad -\alpha\beta \cos \phi = g_{12} = g_{21} \tag{4}$$

and replacing c^2 by ds^2, we may write (3) in the generalized *fundamental metrical* form

$$ds^2 = g_{\mu\nu} dx_\mu dx_\nu \qquad \mu, \nu = 1, 2 \tag{5}$$

If we let μ and ν take on the values 1 and 2 independently of each other, we have four terms, which together with (4) give us expression (3). Therefore (5) is a sum[1] consisting of four terms.

The expression ds is commonly referred to as "the line element of the plane." The numbers $g_{\mu\nu}$ indicate how, at a given place, the length of the line element is to be calculated from the coordinate differentials. Since the $g_{\mu\nu}$ are functions of the position, they are to be written as $g_{\mu\nu}(x_1, x_2)$. These functions determine the metric. If the $g_{\mu\nu}$ are given for every point of the plane, we can calculate the length Δs for any

[1] Note again the rule mentioned on p. 172 that every index occurring twice is to be summed.

curvilinear or straight-line segment, if we know the coordinates of all the points through which the line passes. This length is equal to the sum of the many *ds's* which may be placed along the line, and therefore

$$\Delta s = \int_P^Q ds = \int_P^Q \sqrt{g_{\mu\nu} dx_\mu dx_\nu} \tag{6}$$

This mathematical treatment has the following significance. It divides the function of the customary orthogonal coordinate system into two parts which are quite different. *It leaves the topological function of numbering to the coordinate system, but assigns the metrical function of measuring lengths to the metrical coefficients* $g_{\mu\nu}$. In this way the mathematical treatment divides the description of a plane into a topological and a metrical part and makes possible the analytic treatment of the metric.

This mathematical achievement is of greatest significance for the epistemological problem of space. The method shows that in addition to the metrical function usually associated with the "coordinates of the graph paper," the coordinate system has an important topological function. It consists in the assignment of identification numbers to every point, which can of course be accomplished by curvilinear as well as by rectilinear coordinates. In spite of the arbitrariness of the numbering, something very important is thus achieved, since the numbering determines the mutual neighborhood relations. If three points *A*, *B*, and *C* lie along a line so that *B* lies *between* *A* and *C*, this fact is determined by the position of the points and the line, relative to the coordinate system. To give an example: if we know that in a certain street Mr. X lives in No. 37, Mr. Y in No. 45, and Mr. Z in No. 61, we also know that Mr. Y lives between Mr. X and Mr. Z. But from the given information we know nothing about the respective distances between their houses, since the houses may be located at different distances along the street. We speak of the *topological* function of the coordinate system because it determines the order of the *betweenness* relation.

In contrast to this important role of the description of a plane, the metrical function of the $g_{\mu\nu}$ plays a subordinate role. It cannot change the topological foundation determined by the coordinate system; it merely adds to it a metrical superstructure. In this role, however, the form of the $g_{\mu\nu}$ becomes important for what is commonly referred to as the *Gestalt* of the plane. We can understand this best through the following mathematical considerations.

In the example of Fig. 45 we began with a *plane*. This fact is expressed analytically by a property of the $g_{\mu\nu}$, as we shall see when we investigate transformations to other coordinate systems. Let us imagine that a second coordinate system has been introduced and that the position of the new family of lines is given as a function of the old coordinates, such that we can write

$$\begin{aligned} x_1 &= f_1(x'_1, x'_2) \\ x_2 &= f_2(x'_1, x'_2) \end{aligned} \tag{7}$$

where the functions f are completely arbitrary. We now add the restriction that the transition to the new coordinates must not change any of the metrical relations; this transition, therefore, leads only to a different form of description. We must then specify a new system $g'_{\mu\nu}$ of metrical coefficients relative to the new coordinates x'_μ such that the old relations of congruence are preserved. If two line segments are equally long when measured by the old $g_{\mu\nu}$, in the old coordinate system, they must still be equally long when measured by the new $g'_{\mu\nu}$ in the new coordinate system. This requirement leads to the condition that

$$ds^2 = g_{\mu\nu}\, dx_\mu\, dx_\nu = g'_{\sigma\tau}\, dx'_\sigma\, dx'_\tau \tag{8}$$

We can therefore say that ds^2 is an invariant of the transformation, and we can show from (8) how the new $g'_{\mu\nu}$ are to be calculated from the old ones.

This proof is not difficult. It follows from (7) that the differentials transform linearly:

$$dx_\mu = a^\mu_\sigma\, dx'_\sigma \qquad a^\mu_\sigma = \frac{\partial x_\mu}{\partial x'_\sigma} \tag{9}$$

Substituting (9) on the left side of (8), we obtain, using the identity with the right-hand side:

$$g'_{\sigma\tau} = \frac{\partial x_\mu}{\partial x'_\sigma}\frac{\partial x_\nu}{\partial x'_\tau} g_{\mu\nu} \tag{10}$$

(10) is the so-called law of transformation of tensors, and $g_{\mu\nu}$ is therefore called a *tensor*. This term denotes nothing but a magnitude which transforms according to (10) when we go over to a new coordinate system.

We are now able to express analytically the characteristic property of Fig. 45, namely, that it describes the relations of a *plane*. If we had chosen rectilinear orthogonal coordinates, the line element would be

$$ds^2 = dx_1^2 + dx_2^2$$

This means that relative to the normal coordinate system the $g_{\mu\nu}$ satisfy the matrix

$$\begin{matrix} g_{11} & g_{12} \\ g_{21} & g_{22} \end{matrix} = \begin{matrix} 1 & 0 \\ 0 & 1 \end{matrix} \qquad (11)$$

We can now introduce the following requirement: since the surface considered in Fig. 45 is a plane, there must be a transformation (7) that changes the $g_{\mu\nu}$ of the curvilinear coordinates into the normal form (11).

This requirement expresses a property of the system $g_{\mu\nu}$. It can be shown that the system $g_{\mu\nu}$ has this property only because it was constructed originally for a plane. If the construction of Fig. 45 had been carried out on the surface of a sphere, we would have obtained a system $g_{\mu\nu}$ for which there is no transformation into the normal matrix (11). We shall study these relationships in more detail, because the mathematical development can be given easily.

Let us assume that the four functions $g_{\mu\nu}$ are given. How shall we formulate the condition that they are to be transformed into a normal matrix? This requirement obviously amounts to the demand for a coordinate transformation (7) satisfying the equations:

$$\frac{\partial x_\mu}{\partial x'_\sigma}\frac{\partial x_\nu}{\partial x'_\tau} g_{\mu\nu} = g'_{\sigma\tau} = \begin{cases} 1 & 0 \\ 0 & 1 \end{cases} \qquad (12)$$

Expression (12) represents four equations for the four partial derivatives of the functions in (7). The number of variables corresponds to the number of equations, but since the variables are *functions*, not *numbers*, we must add the Cauchy-Riemann conditions of integrability

$$\frac{\partial^2 x_\mu}{\partial x'_\sigma \partial x'_\tau} = \frac{\partial^2 x_\mu}{\partial x'_\tau \partial x'_\sigma} \qquad (13)$$

Expression (13) represents four more equations, and we have now eight equations for the four partial derivatives $\frac{\partial x_\mu}{\partial x'_\sigma}$. Because of this over-determination, equations (12) and (13) cannot generally be solved.

We may restate our results as follows: any given system $g_{\mu\nu}$ can be transformed into another system $g'_{\sigma\tau}$ by means of (7). Transformations of this kind, starting with a definite set $g_{\mu\nu}$, do not give us all conceivable systems, however, but merely a limited class. All systems of this class are geometrically equivalent to the initial $g_{\mu\nu}$, and the class as a whole characterizes a definite geometry. Other classes, similarly constructed, would characterize another geometry. A special class is the class which contains the normal system (11); it is the class of Euclidean geometry.

The question now arises whether there exists a special characteristic of the class of Euclidean geometry. Mathematicians have shown that it is possible to formulate such a criterion. For this purpose one has to form a certain mathematical combination of the $g_{\mu\nu}$, $\frac{\partial g_{\mu\nu}}{\partial x_\sigma}$, and $\frac{\partial^2 g_{\mu\nu}}{\partial x_\sigma \partial x_\tau}$, which is called $R_{\mu\nu\sigma\tau}$ and which transforms according to rules analogous to (10). $R_{\mu\nu\sigma\tau}$ is therefore a tensor of rank 4. We can recognize directly from (10) a very important property of all tensors, namely, that if all components of a tensor are zero in *one* coordinate system, they will all be zero in *every* coordinate system. This result follows immediately because all terms disappear in the summation of expression (10). Since it can be shown that $R_{\mu\nu\sigma\tau}$ vanishes for the normal system, it follows that *every* system of the Euclidean class is characterized by the condition

$$R_{\mu\nu\sigma\tau} = 0 \tag{14}$$

$R_{\mu\nu\sigma\tau}$ is called the Riemannian curvature tensor. It is a measure of curvature.[1]

One thing can easily be shown: if we restrict ourselves to the requirement that the $g_{\mu\nu}$ shall assume the normal form (11) for one single point, this requirement can always be satisfied by a suitable choice of the coordinate system. The integrability conditions disappear for this limited requirement, since in this case it is no longer the *functions* $\frac{\partial x_\mu}{\partial x'_\sigma}$ but their *numerical values* at a specific point, which are subject to conditions (12). Thus there exists no longer an overdetermination. However, the normal system cannot be realized for all points by means of the *same* transformation; if we extend a coordinate system that satisfies the conditions for one point, generally it will not satisfy these conditions for the *other* points. In the general case, the system of the $g_{\mu\nu}$ can therefore be transformed only in infinitesimal domains into form (11) of the system of the plane. In this way we formulate analytically the property of curved surfaces, namely, that their infinitesimal elements can be treated as planes, just as an infinitesimal element of a curve can be conceived as a straight line. The shape of curved

[1] In the two-dimensional case, the tensor $R_{\mu\nu\sigma\tau}$ can be replaced by a single number R, the so-called Gaussian measure of curvature.

structures differs from the shape of straight structures only in large dimensions, while their infinitesimal elements have the same form. This intuitively evident rule finds its corresponding expression in the analytic representation of the shape of the surface. Conversely, the formulation of geometry with the aid of metrical functions $g_{\mu\nu}$ may be characterized as the extension of the concept of geometry to such cases in which Euclidean geometry holds only in infinitesimal domains.

Because of this analytic treatment, it is now relatively easy to extend our considerations to three-dimensional space or even to a manifold of an arbitrary number of dimensions. For three dimensions, we only have to vary the index, in our previous formulae, from 1 to 3 in order to achieve an analytic treatment of space. We owe this method to the genius of Riemann, who generalized the Gaussian treatment of surfaces. We may refer here to § 2 for greater detail. The occurrence of the normal form of the $g_{\mu\nu}$ in the infinitesimal domain means, for the three-dimensional case, that every Riemannian space is Euclidean in infinitesimal domains. Curved space is thus composed of plane differential elements, just like curved surfaces.

To deal with the four-dimensional space-time manifold we require a minor change, since its metric is of the indefinite type, as we have shown in § 28 and § 29. The normal form of the $g_{\mu\nu}$ then becomes

$$\overline{g_{\mu\nu}} = \begin{bmatrix} 1 & 0 & 0 & 0 \\ 0 & 1 & 0 & 0 \\ 0 & 0 & 1 & 0 \\ 0 & 0 & 0 & -1 \end{bmatrix} \tag{15}$$

However, this makes hardly any difference in the analytic treatment. The values in infinitesimal domains now assume, of course, the indefinite form (15). We can comprise all these formulae by one expression, varying the index from 1 to 4:

$$ds^2 = g_{\mu\nu}\, dx_\mu\, dx_\nu \qquad \mu, \nu, \sigma, \tau = 1 \ldots 4 \tag{16}$$

$$x_\mu = x_\mu(x'_1 \ldots x'_4) \tag{17}$$

$$(18) \qquad dx_\mu = \frac{\partial x_\mu}{\partial x'_\sigma} dx'_\sigma \qquad\qquad g'_{\sigma\tau} = \frac{\partial x_\mu}{\partial x'_\sigma} \frac{\partial x_\nu}{\partial x'_\tau} g_{\mu\nu} \qquad (19)$$

How are we to ascertain the metric of the four-dimensional space-time manifold by means of physical measurements? For this purpose we need:

1. A coordinate system, i.e., all point events must be numbered.
2. The 10 metrical functions $g_{\mu\nu}$ relative to this coordinate system.

We must now consider this problem from the physical point of view. For the time being we shall not study the first of these two tasks, but shall simply assume that such a numbering is given. Pursuing the second task, however, we shall be led into the theory of gravitation.

§ 40. GRAVITATION AND GEOMETRY

We ended our discussion of the problem of gravitation in § 37 with the statement that the special theory of relativity can apply only in infinitesimal domains. We have previously characterized the space-time relations of the special theory of relativity by the fact that the geometry defined by clocks, measuring rods, and light assumes the indefinite type metric in the special form of (1, § 29). In the general case this geometry holds only in infinitesimal domains; a more general geometry, of the kind developed in the previous section and characterized by a system of $g_{\mu\nu}$, holds for gravitational fields.

In order to carry through this idea, let us recall the example of the curved surface. If human beings live on a curved surface the geometry of which they want to determine, they must first introduce a system of coordinates. Their choice of a coordinate system is not limited by restrictions of a metrical kind. Let us imagine that they carry around infinitesimal measuring rods which they use to define the relation of congruence on the entire surface. If they place a unit rod at a point P, then its ends will determine a certain coordinate difference dx_μ. If they now combine the dx_μ according to the special formula

$$dx_1{}^2 + dx_2{}^2 = ds^2$$

then the resulting ds^2 is by no means equal to 1. They would first have to introduce "correction factors" $g_{\mu\nu}$, such that

$$g_{\mu\nu} dx_\mu dx_\nu = ds^2 \tag{1}$$

becomes equal to 1. This is very easily accomplished; many different choices of the numerical factors lead to this result. If we require, however, that ds^2 be equal to 1 for the *same* $g_{\mu\nu}$ and any direction of the rod rotated in P, the $g_{\mu\nu}$ will be determined uniquely. We might imagine that the $g_{\mu\nu}$ are found by experiment. The numbers $g_{\mu\nu}$ of the individual points can then be combined into functions.

It is easily recognized that this procedure results in an over-determination. If the rod is placed in the direction of the line $x_2 = \text{const.}$, then $dx_2 = 0$ and (1) reduces to $g_{11} dx_1{}^2$. Since this expression is to be equal to 1, it will determine g_{11}. In the corresponding fashion we find g_{22}, by placing the rod in the direction of line $x_1 = \text{const.}$ If we add one slanted direction, this experiment will determine $g_{12} = g_{21}$. It must therefore be an axiom, i.e., a matter of truth or falsehood, that the $g_{\mu\nu}$ thus obtained will satisfy the condition $ds^2 = 1$ for all positions of the unit rod. This axiom must of course be considered an empirical law. If it were not confirmed, we would be left with two alternatives. The first of these makes use of the fact that the axiom is satisfied only in infinitesimal domains and that therefore every procedure using finite rods is liable to contain an error. We would assume, therefore, that the unit rod was not chosen sufficiently small and would claim that the axiom would be satisfied if the rod were shortened. The $g_{\mu\nu}$ would then have to be treated within the original point region, not as constants, but as functions of the position, and the original length of the rod would be calculated not as ds, but as $\int ds$. Whether this assumption is correct can be decided, since it is possible to test whether the hypothesis is better satisfied as the rod becomes shorter. We shall then infer inductively whether the axiom can be maintained in the limiting case. If it should turn out that the described method is inadequate, a second alternative would be left. We would have to repeat the above process with line elements of higher order, since line elements of second order did not accomplish the task. We might start first with line elements of the fourth order

$$ds^4 = g_{\mu\nu\sigma\tau} dx_\mu dx_\nu dx_\sigma dx_\tau \tag{2}$$

We would thus have a greater number of metrical coefficients available and we could adjust them to a greater number of conditions. The higher the order of the line element, the more adaptable it will be to the actual conditions in nature. The fact that line elements of the second order prove to be sufficient indicates a property of reality. It actually says nothing more than that measuring instruments obey Euclidean geometry in infinitesimal domains.

Here again an empirical statement is to be added to a coordinative definition to guarantee that the definition be unique. However, this fact does not deprive the coordinative definition of its definitional character; rather, the determination of the $g_{\mu\nu}$ for a given coordinate system is a physically meaningful and definite task only if the coordinative definition of congruence has previously been given. We now have to give an analogous development for the space-time manifold.

The coordinative definition of congruence is again given in terms of clocks, rods, and light, just as in the special theory of relativity. The interval ds^2, realized by these measuring instruments, will be defined as the unit and after a coordinate system has been chosen, the corresponding $g_{\mu\nu}$ are calculated. We thus construct a network of infinitesimal unit widths. The clocks supply $ds^2 = -1$, the measuring rods $ds^2 = +1$, and light $ds^2 = 0$. Through experiment we discover at every point those numbers $g_{\mu\nu}$ by which the coordinate differentials must be multiplied in order that the interval will equal 1. The numbers $g_{\mu\nu}$ are again combined into functions.

This method is the same as that used in the special theory of relativity, where we constructed the metric by means of a coordinative definition based on clocks, measuring rods, and light. However, there the coordinates were chosen in such a way that the $g_{\mu\nu}$ satisfied the normal matrix (15, § 39). We showed in § 27 how this choice of the special coordinates is to be carried out. If we now admit arbitrary coordinate systems, the normal matrix will no longer suffice, and the more general form (1) must be employed.

Obviously, more than a definition is involved in this procedure. As before, the system of the $g_{\mu\nu}$ for one point is determined by a certain number (10) of positions of the interval; that the other positions introduce no contradictions must be stated as an additional axiom. Since the slanted positions of the interval are equivalent, in the geometrical interpretation, to different *states of motion* of the measuring instruments (see § 29), the axiom constitutes an assumption about the behavior of moving measuring instruments. The axiom is already partially expressed in the special theory of relativity, where it is formulated in the matter-axioms and asserts that the measuring instruments of uniformly moving systems adjust to the relativistic light-geometry. In the general theory we must extend this axiom to include accelerated measuring instruments. This extended formulation states that the metrical behavior of a body depends only on its velocity, not on its acceleration. If, for instance, a clock in accelerated motion has the same velocity as a uniformly moving clock at the precise moment when the two pass each other, then the two clocks will show the same flow of time during this common infinitesimal period of time.

Whereas we have so far dealt only with the determination of the space-time geometry of gravitational fields, we shall now investigate an additional feature of the theory. It is asserted that the metrical tensor $g_{\mu\nu}$ is identical with the tensor of the gravitational field which we found to be necessary for the characterization of gravitation (cf. § 37). This conclusion follows directly from the principle of equivalence.

If we establish a metric at one point in a local inertial system K', the $g'_{\sigma\tau}$ will satisfy there the normal matrix. If we now describe the same local world region from an accelerated system K, the $g_{\mu\nu}$ of this system can no longer satisfy the normal matrix, since they will result from the $g'_{\sigma\tau}$ by a transformation which does not belong to the Lorentz transformations and which will therefore destroy the orthogonal form (1, § 29) of the line element. The $g_{\mu\nu}$ will therefore characterize the state of acceleration of K. If they characterize the state of acceleration of K, however, they must also characterize the gravitational field which exists for K because of the principle of equivalence. The metrical tensor must therefore also be the gravitational tensor.

We may clarify this relation in the following manner. If we go from an inertial system to an accelerated system by means of the transformation (17, § 39), the coefficients of the equation of the transformation will contain the acceleration a. This magnitude enters through the $\frac{\partial x'_\sigma}{\partial x_\mu}$ into the $g_{\mu\nu}$, and the $g_{\mu\nu}$ contain therefore the acceleration a of the chosen coordinate system. And since, conversely, a can also be interpreted as the intensity g of the gravitational field, the $g_{\mu\nu}$ contain also this magnitude, and are therefore a measure of the gravitational state existing for this coordinate system. In a simple form, this idea is found even in the Newtonian theory: because of the identity of gravitational and inertial mass, we can identify the gravitational field g with the acceleration of freely falling bodies. Therefore it is unnecessary to distinguish between the two magnitudes, which originally were conceptually distinct.

One might now ask whether this new assertion contains anything factual. Does gravitation mean more than what is expressed by the metrical function of the tensors $g_{\mu\nu}$? It does indeed. Gravitation can also be recognized by the appearance of forces, by the bending of elastic beams, and the motion of mass points (planets). The factual import of the new claim lies in the fact that the same physical magnitudes that determine the motion of mass points and the bending of beams also determine the length of rods, the period of clocks, and the path of light. This assertion is, of course, of extraordinary physical importance. Since we combine the behavior of measuring instruments under the concept of metric, and the appearance of forces of the kind mentioned under the concept of gravitation, we may formulate the new assertion as the *identification of metrical and gravitational field* and refer to the field that pervades the entire space as the *metrical field.* This identification is a consequence of the principle of equivalence.

We must now point out a difficulty closely connected with this result. The identification just mentioned is based primarily on the fact that the geometry of the theory of relativity includes the dimension of time. If we interpret the $g_{\mu\nu}$ which result from the transition from a normal to another system, as the gravitational field, we think of a transformation of the state of motion. Since transformations of the state of motion are not the only coordinate transformations of four-dimensional space-time manifolds, however, the identity of the metrical tensor and the gravitational tensor says more than was originally expressed by the principle of equivalence. Even purely spatial coordinate transformations must now be interpreted as transformations to new gravitational fields. If we change to three-dimensional polar coordinates, for example, while the time coordinate remains unchanged, the $g_{\mu\nu}$ will assume a form different from (15, § 39).[1] For these coordinates there must therefore exist a gravitational field. Pure time transformations which leave the space coordinates unchanged and therefore do not constitute a change in the state of motion but stand for a redefinition of simultaneity, e.g., the definition of simultaneity given by (1) in § 27, will also cause a deviation of the $g_{\mu\nu}$ from the normal form and produce a gravitational field. Through the identification of the $g_{\mu\nu}$ system with the gravitational field the concept of gravitation receives another, though inessential, extension which goes even beyond the introduction of dynamic gravitational fields. But in virtue of this comprehensiveness the concept of gravitation becomes accessible to the mathematical treatment by means of Riemannian geometry.

If we wish to avoid this all too general concept of gravitation, we may use the concept of the metrical field. All $g'_{\sigma\tau}$-systems derived from a $g_{\mu\nu}$-system by means of coordinate transformations are merely different resolutions of the same tensor into different sets of components. This tensor, the metrical field, is therefore independent of specific

[1] Setting $r = x_1$, $\phi = x_2$ and $\theta = x_3$, the $g_{\mu\nu}$ become:

$$g_{\mu\nu} = \begin{matrix} 1 & 0 & 0 \\ 0 & x_1^2 & 0 \\ 0 & 0 & x_1^2 \cos^2 x_2 \end{matrix}$$

It can easily be seen that the partial derivatives $\frac{\partial g_{\mu\nu}}{\partial x_\sigma}$ do not vanish throughout.

coordinate systems. If we now identify the gravitational field with the metrical field g, polar coordinates and systems that have a differently defined simultaneity will have the same gravitational fields as the system from which they resulted through transformation. We can thus retain the intuitively plausible property of the gravitational field, namely, its independence of such coordinate transformations. This statement means, of course, that we will also have to accept the consequence that transformations of the state of motion will not change the gravitational field either, since they too leave the metrical field invariant. All these difficulties can be avoided if we remember that the earlier concept of gravitation is now divided into two separate concepts. One of these is the metrical field, which has taken over from the earlier concept of gravitation the property of being independent of the coordinate system; the other is the actual system of components of the metrical field, which has taken over the remaining properties of the earlier concept of gravitation and which is therefore commonly referred to as the gravitational field. We should not be surprised to find that this narrower concept of gravitation refers to fields that cannot properly be subsumed under the earlier concept of gravitation.

We have to pursue our analysis further. Since we are able to show that the metrical field of space is at once a manifestation of gravitation, there arises the possibility of asking for a *cause* of the metrical field. It has not been customary to ask this question, because the geometry of space has commonly been accepted as a fact requiring no causal explanation. A cause of gravitation was known, however. Ever since Newton, gravitation has been looked upon as the effect of masses. To this conception was added Mach's idea that the masses are also the cause of inertia. If gravitation and inertia are now combined to form the field $g_{\mu\nu}$, one must conclude that the cause of the field $g_{\mu\nu}$, and therefore also the cause of geometry, is to be found in the distribution of masses, and that there must be a law of nature stating how the $g_{\mu\nu}$ field is related to the distribution of matter.

In classical mechanics this law is given by Newton's equation

$$f = \frac{m_1 m_2}{r^2} \tag{3}$$

or the corresponding differential law

$$\Delta\phi = 2\pi\rho \tag{4}$$

the so-called Poisson equation. It states how the matter ρ determines the gravitational field ϕ. In Einstein's theory of gravitation, the tensor field $g_{\mu\nu}$ takes the place of the scalar field ϕ, and instead of the scalar mass ρ we have the matter tensor $T_{\mu\nu}$ which represents the density of mass as well as the internal tension of matter. Guided by this analogy, Einstein guessed, so to speak, at the new law:

$$R_{\mu\nu}-\tfrac{1}{2}g_{\mu\nu}R = -\kappa T_{\mu\nu} \qquad (5)$$

On the left hand side of (5) we find a complicated function of the $g_{\mu\nu}$, which is abbreviated by the symbols "$R_{\mu\nu}$" and "R"; κ is a constant. This law implies (4): it states how the gravitational state $g_{\mu\nu}$ is to be calculated from the distribution of matter $T_{\mu\nu}$. Since $g_{\mu\nu}$, however, also represents the metrical tensor in contrast to the classical gravitational function ϕ, (5) states something essentially new compared to (4): *it states how the geometry of the universe is determined by the distribution of matter.* Einstein's field equations assert therefore a new law of nature, the existence of which was not suspected before.

Equation (5) represents the most fundamental idea of Einstein's theory of gravitation in the physical as well as the philosophical sense. Let us first study its physical significance. Equation (5) is the key to the previously mentioned relativity effects: the bending of light and the retardation of clocks in a gravitational field. Since the behavior of clocks and light is determined by the $g_{\mu\nu}$, and since $T_{\mu\nu}$ represents the masses of the stars, these effects can now be calculated rigorously from (5). The considerations outlined in § 36, based on the Newtonian theory of gravitation, can at best be approximations. Furthermore, expression (5) implies a change in the physical content of the law of gravitation. It differs from (4) even if we consider the $g_{\mu\nu}$ only as gravitational functions. This is evident, since in (5) the gravitational field at each point is characterized by 10 parameters, whereas in (4) only by one. Relation (5) can therefore contain the Newtonian gravitational field ϕ only as an approximation. It can indeed be shown that this approximation is realized when all components of $g_{\mu\nu}$, with the exception of g_{44}, approximate the normal form, in which case $g_{44} = \phi$. If we were to calculate from (5) a purely gravitational process, e.g., the motion of a mass point around a mass center, the result would differ slightly from that calculated from (4). Einstein's planetary motion is therefore different from that of Newton; it asserts a slight rotation of the elliptic orbit in addition to the elliptic motion. This consequence is strikingly confirmed by the advance of the perihelion of the planet Mercury, a phenomenon that has been known for a long time.

Let us now turn to the question of what is new in the identification of gravitation and metric from the philosophical point of view. It has occasionally been said that this conception deprives gravitation of its physical character and that gravitation therefore becomes geometry. We shall now investigate whether this conclusion is justified.

We have learned in § 6 about the difference between *universal* and *differential* forces. These concepts have a bearing upon this problem because we find that gravitation is a universal force. It does indeed affect all bodies in the same manner. This is the physical significance of the equality of gravitational and inertial mass. If gravitational and inertial mass were not equal, we would not be able to look upon the paths of freely falling mass points as (four-dimensional) geodesics, since different geometries would result for the various materials of the mass points. Furthermore, due to the influence of gravitation upon light, light may be regarded, in this geometry, as a realization of $ds^2 = 0$, and the gravitational effect on infinitesimal clocks and rods is such that they may be considered as realizations of $ds^2 = \pm 1$. This universal effect of gravitation on all kinds of measuring instruments defines therefore a single geometry. In this respect we may say that gravitation is *geometrized.* We do not speak of a change produced by the gravitational field in the measuring instruments, but regard the measuring instruments as "free from deforming forces" in spite of the gravitational effects.

However, we have seen that for geometry, as for all other phenomena, we must pose the question of causation. Even if we do not introduce a force to explain the *deviation* of a measuring instrument from some normal geometry, we must still invoke a force as a cause for the fact that *there is a general correspondence of all measuring instruments.* We have expressed this idea in § 31 by means of the concept of *adjustment.* In this sense we must ascribe to the gravitational field the physical reality of a force-field. We regard this force-field as the cause of geometry itself, not as the cause of the disturbance of geometrical relations. We can even demonstrate physically the existence of this force-field: since the gravitational field is measured by the same measuring instruments as those used for geometry, these measuring instruments are at once indicators of the gravitational field. We are therefore reversing the actual relationship if we speak of a reduction of mechanics to geometry: *it is not the theory of gravitation that becomes geometry, but it is geometry that becomes an expression of the gravitational field.* The theory of relativity did not convert a part of physics into geometry. On the contrary, even more physics is involved in geometry than was suggested by the empirical theory of physical geometry: the geometry of the universe is not only a fact that can be *ascertained empirically*, but also a fact to be *explained by the effects of forces*. In addition to the problem of the *measurement* of physical space, known since Gauss, Riemann and Helmholtz, Einstein introduced the problem of a *scientific explanation* of physical geometry, which finds its mathematical solution in the gravitational field equations.

According to Einstein's theory, we may consider the effect of gravitational fields on measuring instruments to be of the same type as all known effects of forces. This conception stands in contrast to the view which interprets the geometrization of physics as an exclusion of forces from the explanation of planetary motion. This view is based on the laws of motion of a mass point. In classical mechanics these laws state that a force-free point travels in a straight line. Einstein substitutes "the straightest line" for "a straight line." Since, however, the characterization of the gravitational field is contained in geometry, these laws include even the case when the point is subject to gravitational forces. Planetary motion and the motion of a force-free point are thus combined into a single law, which states that a mass point moves along the geodesic. This result suggests a purely geometrical conception of gravitational motion. Accordingly, the planet does not follow its curved path because it is acted upon by a force, but because the space-time manifold leaves it, so to speak, no alternative path. Its motion resembles that of a sphere rolling on an irregular surface along some definite curves.

This view is correct, inasmuch as we have to consider, in accord with the principle of action by contact, the planetary action to be caused by the state of the metrical field in the immediate neighborhood of the planet, thus relinquishing the Newtonian concept of action at a distance, which moves the planet around the sun as though by a string. On the other hand, not only is the metrical field determined [1] by distant masses through a genuine physical process of causal propagation, but the effect of the metrical field itself on the planet may be interpreted as a genuine physical force guiding it on its path. A causal determination of the orbit by the nature of space is not compatible with our customary physical concepts. If a sphere rolls on a material plane, it is only a schematization to say that the geometry of the plane

[1] Changes in the metrical field due to changes in the distribution of matter propagate only with the speed of light, not instantaneously.

influences the path of the sphere. We know that a more detailed investigation would reveal the presence of molecular force-fields, which affect the molecules on the surface of the sphere and thus force it into a definite path. Geometrical effects reduce therefore, even in classical physics, to dynamic effects. Even in empty Euclidean space, in which a mass point travels in a straight line according to the law of inertia, we must not think of this path as an effect of the geometry that permits the mass point to travel on the shortest path. Even here the guiding gravitational field, whose tensor $g_{\mu\nu}$ is reduced to the normal form, manifests itself and, due to its force, compels the mass point to travel along a definite orbit.

The combination of gravitation and geometry therefore does not force us to renounce dynamical conceptions, but teaches us that such conceptions are applicable even to those cases treated previously in a purely geometric fashion. It is no longer possible to assume a geometry of space independent of physical realities. *Geometric measurement is a handling of indicators; therefore, the metric of the indicators is at once the measure of the field that determines their adjustment.*

§ 41. SPACE AND TIME IN SPECIAL GRAVITATIONAL FIELDS

We shall now give a more detailed description of the geometry of the gravitational field. For this purpose we refer to § 27, where we characterized gravitation-free space by certain specializing axioms that permitted the choice of selected coordinate systems. It turns out that the construction in § 27, which starts with the elementary concept of time order and leads to the complete metric, is of fundamental importance: the levels of this construction are levels of increasing generalization of the gravitational field if we proceed in the opposite direction and gradually omit the specializing axioms. The space-time properties of the gravitational field at each level are thus given by those axioms which are still satisfied on that level.

The first step toward this generalization consists in the omission of axiom V,[1] which postulates the possibility of the choice of a coordinate system with *Euclidean* light-geometry. The space characterized by the remaining four axioms (I–IV) corresponds to the space of the *static gravitational field.* Again, as in the case of the gravitation-free

[1] [The numbers refer to the statement of the axioms in A.—M.R.]

space, we must add the condition that it must be possible to connect the space points of the coordinate system by rigid rods. The light-axioms alone will again be insufficient to give an exhaustive determination of the state of motion. In the language of the Riemannian geometry, the static gravitational field is characterized by the fact that the $g_{\mu\nu}$ no longer assume the normal form $\overline{g_{\mu\nu}}$ (15, § 38), but are still of a rather simple kind because they are given by functions of the space coordinates alone, independent of time, with the additional condition that the $g_{\alpha 4}$ ($\alpha = 1, 2, 3$) vanish.

In such a gravitational field, space and time still have very simple properties, like those of the gravitation-free case except that the space-geometry is non-Euclidean. Time, on the other hand, retains the special property that Einstein's definition of simultaneity (2, § 19, with $\epsilon = \frac{1}{2}$) leads to a symmetrical and transitive synchronization, as we showed in § 27 prior to the introduction of axiom V. We may therefore understand by the time of a static gravitational field the same as we did by the time of an inertial system. A difference exists only in that transported clocks deviate not only relative to the *simultaneity* of the system but also relative to the *unit of time* transmitted by the light signal. The unit of time can therefore no longer be transported, yet the difference between the transported unit and the projected unit depends only on position and not on the path of transport.

This statement is a consequence of the fact that the clock indicates the interval, which reduces to

$$-g_{44}dx_4{}^2 = ds^2 \tag{1}$$

for a clock at rest. If g_{44} is different at the two points P and P', then the corresponding dx_4 must also be different, since the unit clock supplies the same ds. If we now call dx_4 the unit of the coordinate time at P, and dx'_4 the time unit at P' which was transmitted by means of light signals from P according to (1, § 27), we then have

$$dx_4 : dx'_4 = \sqrt{g'_{44}} : \sqrt{g_{44}} \tag{2}$$

This formulation is a rigorous derivation of the retardation of clocks in gravitational fields (the red shift), which was mentioned briefly in § 36. It states merely that the unit of the time *coordinate* cannot be transported. The *measure* of a time interval at P is not given by dx_4, however, but by $\sqrt{g_{44}}\, dx_4$; and since this measure is everywhere realized by clocks, *the measure of a time interval* is obviously transportable. This peculiarity is more clearly characterized in a different fashion. One might try to define a time coordinate such that equal measures of time would correspond everywhere to equal dx_4, i.e., g_{44} would become equal to -1. Though we can at first combine this rule for the measure of time with an arbitrary definition of simultaneity which satisfies the inequality for ϵ given in (3, § 19), simultaneity would gradually shift in such a manner that later this inequality would be violated and events would be simultaneous that are not indeterminate as to their time order. The time thus introduced would therefore violate the basic topological requirement of time measurement. We may therefore say that it is impossible to define in a gravitational field a time *coordinate* that corresponds everywhere directly to the *measure* of time.

Let us now proceed to measurements of space. Spatial congruence is given by the characteristic length of rigid rods. Since the events that are simultaneous at both ends of the rod in the sense of the natural length are also simultaneous in the sense of the coordinate time in a static gravitational field, we have for the characteristic length $dx_4 = 0$, and ds^2 reduces to

$$d\sigma^2 = g_{\alpha\beta}dx_\alpha dx_\beta \qquad \alpha, \beta = 1, 2, 3 \tag{3}$$

This means that spatial congruence is defined directly by the rigid rod. Two rods equal in length when compared at one place are also called equally long when located at different places. As in the case of the time unit of the clock, this definition of congruence applies only to the *measure* of the segment determined by the rods, while the respective spatial coordinate differentials dx_α ($\alpha = 1, 2, 3$) are everywhere different. It is impossible, therefore, to construct a spatial coordinate system that eliminates this difference between the rod and the coordinate differentials throughout.

Space measurements and time measurements behave therefore in a parallel fashion. Measuring instruments still supply the *measure* of length, but they can no longer be used to define a uniform coordinate system. In gravitation-free fields the two concepts coincide; there, clocks and rods supply a measure as well as a normal coordinate system. In the gravitational field we can no longer construct a normal coordinate system. Instead, the measuring instruments supply the $g_{\mu\nu}$ that correspond to the chosen coordinate system.

These relations have a peculiar consequence regarding the velocity of light. Let us imagine the Michelson experiment (see Fig. 29) to be performed in a gravitational field, with the arms of the apparatus extended so far that they go beyond the region in which the special theory of relativity applies with a sufficient degree of approximation. The two arms are to be equally long in the sense of (3), i.e., when small measuring rods are placed along the arms, they mark off an equal number of segments on both arms. It can be shown that then the round-trip times of the light signals along the two arms become unequal, thus

$$\overline{ABA} \neq \overline{ACA} \tag{4}$$

Therefore the matter-geometrical equality of distances no longer coincides with the light-geometrical equality. (The two expressions in (4) would be equal only if the arms of the Michelson apparatus became infinitesimal.) In gravitational fields, the velocity of light is no longer constant, and light no longer travels along the shortest path in space. In a static gravitational field however, the velocity of light is still independent of both time and the direction in which a given distance is traveled. We cannot say, of course, that the light-geometrical definition of the equality of length (10, § 27), which would make the velocity of light constant even in this case, is incorrect. Since the light-axioms I–IV, 2 are still satisfied, the light-geometrical definition can be carried through consistently in the static gravitational field, but it would contradict the matter-axioms. The ds^2 defined by the transport of clocks and rods would no longer agree with that defined by light. The choice of the definition is arbitrary. Einstein prefers the matter-geometrical definition for reasons of descriptive simplicity.

These geometrical relations supply a good example of a case where the "natural" definitions of the spatial metric do not always lead to the *same* metric. The light-geometrical definition of length presents a standard just as simple and natural as that given by the matter-geometrical one. Whereas the two coincide in the gravitation-free field, they no longer do so in a gravitational field. We shall find a similar discrepancy upon further generalization.

Let us now take the second step in our generalization by omitting the round-trip axiom IV, 2, according to which there exist coordinate systems in which the time needed to travel around a triangle is equal in both directions. The space characterized by the remaining light-axioms corresponds to the *stationary* gravitational field. Again we must add the condition that it be possible to connect all space points of the coordinate system by rigid rods, in order to obtain a sufficient specification of the coordinate system. In the language of Riemannian geometry, the stationary gravitational field is given by the condition that the $g_{\mu\nu}$ are still independent of time, but that in contrast to the static field, the $g_{\alpha 4}$ ($\alpha = 1, 2, 3$) do not vanish.

The latter property has the consequence that time determination becomes more complicated. The velocity of light is now different along the two directions of a given segment. We can recognize this fact from the failure of the round-trip axiom, since it is only this axiom that renders transitive the synchronization constructed according to Einstein's definition. If the axiom is no longer satisfied, then the definition of synchronization will not be satisfied for two clocks each set relative to a central clock (using $\epsilon = \frac{1}{2}$). This result means that there is no definition of time according to which it is possible for the velocity of light everywhere to be equal in both directions along the same segment.

An example of a stationary system is given by a circular disk that rotates relative to an inertial system K. For this disk, we can obtain a rather simple definition of time if we set up a clock at its center and synchronize from there all clocks on the disk, with $\epsilon = \frac{1}{2}$ (1, § 27). Two clocks on the perimeter of the disk will then have relative to each other a different ϵ which, however, is constant in time and depends on the respective positions of the clocks. This definition of time is therefore quite plausible. It is, incidentally, identical with the Einsteinian definition of time in K, i.e., every clock on the disk always shows exactly the same time as that particular clock of K over which it happens to be located at that moment.

It is important to note that in stationary systems even space measurements assume some very complicated properties. The characteristic length ds of the unit rod, which represents an *interval*, is no longer identical with the spatial length $d\sigma$ of the rod according to (3), because the length $d\sigma$ is measured by two events that take place at the two ends of the rod and are simultaneous in the sense of the simultaneity of stationary systems. However, since this simultaneity does not agree with Einstein's $\epsilon = \frac{1}{2}$, the events are not simultaneous in the sense of the the characteristic length. This fact leads to the consequence that the spatial congruence of stationary systems is no longer defined by rigid rods. A correction factor [1] dependent on both position and direction must be included. In contrast to the static field this result means not only a difference in the spatial coordinate differentials determined by the rod but also a difference in the measure of length determined by the rod.

[1] This correction is connected with the so-called circle paradox. See A., § 44.

Two rods equally long when compared at one place are no longer equally long when they are apart (although they would always be found to be equally long when compared at one place after having been transported along different paths). The spatial geometry of stationary gravitational fields is therefore no longer based on the definition of congruence in terms of rigid bodies.

The variation of the unit in transport can be interpreted as the effect of a force which would then have the property of a universal supplementary force. This definition of congruence seems to be in conflict with our previous considerations of § 5 and § 6, according to which universal forces were excluded by definition. The present situation, however, is more complicated: whereas Einstein's definition of the ds^2 in terms of clocks and rods excludes universal forces from the *four-dimensional* manifold, such forces now are unavoidable in *three-dimensional* space.

If conversely we were to define spatial congruence on the rotating disk by rigid rods under exclusion of universal forces—which is of course also permissible—we would then introduce universal forces into the four-dimensional space-time manifold, and Einstein's realization of the ds^2 would thus have to be renounced. There appears therefore a discrepancy between two natural definitions of the metric within matter-geometry itself. Due to this fact, the transforming away of universal forces is no longer completely in our hands. This result shows that there are limitations to the arbitrariness of definitions.

§ 42. SPACE AND TIME IN GENERAL GRAVITATIONAL FIELDS

Finally, we shall study the most general gravitational fields, omitting axiom IV, 1, which postulates the existence of stationary systems in which the time $\overline{ABA}$ of the path of a light ray is constant for any pair of points. In the language of the Riemannian geometry, these are systems with arbitrary $g_{\mu\nu}$ [1] that change with position and time.

How are we to conceive the spatio-temporal metric of such a manifold? Here we are faced by a world of the παντα ρει, where actually everything is in a state of continuous flux. If we were to measure the

[1] We would have to add certain conditions in terms of a determinant which assert the indefinite character of the fundamental quadratic form. See A., § 48.

diameter and circumference of a circle, their quotient today would not be the same as that of tomorrow. Indeed, if we were to make a material circle of some wire, it would no longer be a circle tomorrow, having been bent by fluctuating gravitational fields. Static structures are therefore impossible in this world, which bends and shifts everything with time. The concept of rigidity loses its meaning in this world, in which only rubber bands would be "solid" bodies that do not break. We recognized earlier (§ 34) the relativity of the concept of relative rest, and we showed that the rigidity of solid bodies is merely a convenient definition of relative rest. Certain coordinate systems have the property that all their points can be connected by rigid rods and that the points thus connected retain their identity through the course of time. These special properties disappear in the most general gravitational fields. We could, of course, still define a framework of iron beams as internally at rest, defining away in this fashion its internal motion. This would be possible so long as the fluctuations of the metrical field were not strong enough to exceed the elastic limit of the material and tear the framework. However, we would also find that a corresponding framework made of copper beams, which had originally been equal in size and shape to the first, would assume a different size and shape after some time. Its points must therefore have moved relative to the first framework. Hence solid bodies would therefore no longer define a unique coordinate system as internally at rest, due to differences in their elastic properties. We thus lose the possibility of defining a geometrical state as rigid in terms of solid bodies, since this definition was based primarily on the uniform behavior of different materials. There is therefore no longer a geometry of rigid bodies.

Similar complications arise for light-geometry. Light rays can no longer be defined as straight lines, because their paths back and forth do not coincide. A straight line from A to B would be different from the straight line from B to A. Even the points through which the light ray travels in one direction from A to B will vary with time. Light also lcses its uniqueness with respect to time, since there is no longer a specific definition of simultaneity on the basis of which the velocity of light is constant. In whatever manner we set our clocks, not only will the velocity of light generally be different at different points but it will also have different values for different line elements through the same point—and even these values are subject to change in the course of time. There is therefore no light-geometry in these most general gravitational fields.

It seems reasonable to ask whether there is anything left which might be called geometry. If there are no rigid bodies, then our previous definition of physical geometry, as the system of the relations of rigid bodies, loses its applicability; and if there is no light-geometry either, then light cannot take the place of the rigid body in the coordinative definition of geometry. The coupling of geometry and gravitation that follows from the theory of relativity has therefore a peculiar consequence. Its greatest success consisted in its *explanation* of geometry, in which it revealed the behavior of measuring instruments as an effect of a gravitational field. But this conception subjects geometry to the variability of gravitational phenomena, and geometry loses its definiteness in fields in which the adjustment of measuring instruments is not uniform. Is it reasonable, then, to speak of a geometry in such gravitational fields?

Let us first see how a physicist solves the problem. He makes use of the applicability of the special theory of relativity to infinitesimal domains. In small regions—which incidentally are quite large with respect to the human organism, since deviations from the geometry of the special theory of relativity appear only in astronomical dimensions—he can determine the metrical relations with sufficient accuracy according to the methods of the special theory of relativity. He can use the light-geometry, and he will find a sufficient number of rigid bodies. He can avail himself of these bodies if he assumes that the axioms are satisfied for limited regions of space only, and, since the $g_{\mu\nu}$ are regarded as variable in time, if he limits himself to small time intervals. He would therefore consider only large-scale phenomena as fluctuating. He might now determine even a large-scale metric by treating the fluctuating phenomena in such a fashion that the metric corresponds everywhere, in infinitesimal domains, to that of the special theory of relativity. He would therefore describe the path of light rays and the motion of the planets in terms of correction factors such that in turn these physical processes would define a metric corresponding everywhere in infinitesimal domains to the special theory of relativity. This method yields a physical geometry for large dimensions, which, of course, does not contain "direct" realizations of the geometrical elements; it would be defined by "indirect connection" to physical reality. We can thus impose upon this flowing world a conceptual system that makes the chaos appear as an *ordered* flow. Geometry would then be that conceptual framework which brings order into the chaos.

This definition of physical geometry is, of course, very abstract. It is constructed by reference to infinitesimal domains, and we must admit that the concepts of space and time seem to be meaningful only for these domains. Its space-time determinations would therefore not express the relations of the behavior of large physical objects, but merely describe the behavior of these physical objects relative to infinitesimal measuring instruments. This conception leads to a considerable limitation of the original meaning of the concept of geometry.

We must therefore ask whether it is perhaps possible to establish between large-scale structures in such a world relations that might be regarded as geometrical. This possibility is suggested by the idea that the relation to infinitesimal measuring instruments will entail certain relations between the large-scale structures themselves. We therefore must try to discover these large-scale relations. The task can indeed be carried out. Mathematically speaking, this problem is the same as asking for the integral properties of a Riemannian geometry which is known to correspond to Euclidean geometry in the differential domain.

The answer to this question is found in the concepts of topology which we have used earlier (§ 12 and § 21). Although the metric of Riemannian spaces is completely arbitrary, their topology is quite definite, inasmuch as the basic topological properties are common to all Riemannian spaces. There may be some variations, of course, in the *topological form*, which may be that of a sphere, a torus, an open space, etc. However, these are merely special cases independent of which there is a common topological basis given by the condition that every Riemannian space must be "plane in its smallest elements," i.e., correspond to the special theory of relativity. It is this topological property that is common to all curved surfaces; all of them are constructed from "infinitesimal planes" and are therefore topologically related. We can formulate these common topological properties, as will be seen presently, by characterizing them as properties of "cut-out regions."

Let us consider a torus. A torus cannot be mapped on a plane topologically (in a one-to-one correspondence and a continuous fashion) because it has a different topological shape. We can, however, cut out of the surface of the torus a region that will become topologically equivalent to a bounded region of the plane, under a suitable choice of a boundary. The curvature does not matter in this case. The impossibility of a one-to-one and continuous mapping applies only to the torus as a whole, not to a cut-out region. Since all curved surfaces behave in this fashion, we can topologically map the cut-out regions of any surface directly upon any other. Thus we have a method to eliminate differences in topological shape: we shall call two manifolds *topologically related* if we can everywhere cut out regions from them which are *topologically equivalent.*

It is important to note that this statement does not employ the concept of infinitesimal domain, since we are not saying that the statement applies to finite regions only in the sense of an approximation and is strictly valid only in the limit. The statement applies rigorously to finite regions. This characterization of common topological properties differs from the formulation referring to "planeness in the smallest elements," or to "Euclidity of the smallest elements," although it is mathematically equivalent to the latter. The latter characterization contains metrical concepts and can therefore be applied only in the limit. The characterization in terms of cut-out regions, however, uses exclusively a topological concept, namely that of one-to-one and continuous mapping, which can be strictly satisfied for non-vanishing regions. If we speak of "cut-out regions," we shall always mean nonvanishing finite regions. One should also note that we merely say that such regions *can be cut out,* since of course not every cut-out region has the described properties. For the sake of convenience we shall adopt the rule that to say "a space has such and such properties in cut-out regions" means that such regions can be cut out.

Our problem is now reduced to finding the topological properties of the cut-out regions in the space-time manifold of the most general gravitational field. The answer to this question is very simple: they are the same topological properties as those displayed by the space-time manifold of the infinitesimal domains. Any curved surface is in its cut-out regions topologically equivalent to the plane. Similarly the space-time order in the cut-out regions of the most general gravitational field must be topologically equivalent to the space-time order of the special theory of relativity. This is the key that opens the door to the characterization of the geometry of the gravitational field.

We have already discussed the topological properties of the space-time order in our analysis of the special theory of relativity. We found there that the concept of time order is of primary importance and that it can be reduced to the concept of the causal chain. The causal chain proved to be the basic topological element of *time order.* The concepts *earlier, later,* and *simultaneous* revealed themselves as ordering concepts by means of which we were able to characterize the most general properties of causal structure. Continued investigation showed furthermore that even *spatial order* can be based on the causal chain. We found there that the definition of spatial distance (§ 27) amounts to saying that a space point is *farther away* when the causal propagation takes *more time* to reach it. When we used these concepts in the light-geometry for the definition of the spatial metric, we first laid down their topological basis in the form of a special coordinative definition. It was the concept of spatial *betweenness* which was thus supplied by the causal propagation. There can be no doubt that this conception reveals the basic meaning of the topology of space. It contains the answer to the question what is *actually meant* by the order of spatially adjacent regions. That we call Sirius very distant and the sun relatively close means nothing other than that a causal chain originating with us will reach Sirius much later than the sun. The "light-year" of the astronomer, which was originally introduced as a matter of expedience, corresponds to the logical archetype of all measurements of length. Time, and through it causality, supplies the measure and the order of space; *not time order alone, but the combined space-time order reveals itself as the ordering schema governing causal chains and thus as the expression of the causal structure of the universe.*

The system of causal ordering relations, independent of any metric, presents therefore the most general type of physical geometry, namely the type realized by large objects even in the most general gravitational fields. If rigidity and uniformity were to disappear, the causal chain would still remain as a type of order. The causal chain is the real process that constitutes the immediate physical correlate of the purely ordering geometry of Riemannian spaces. Although everything is in continuous flux, there is a structure discernible in this flux. It is striated and can be resolved into chains that define a strict topological order. The analysis of the concepts of space and time, which was carried through by the theory of relativity and which has culminated in the denial of any metrical importance of geometry, has clarified the cognitive significance of the concepts of space and time. The order of causal chains is ultimately reflected in all space-time determinations.

The *causal theory of space and time*, to which we were led by the epistemological study of the foundations of space-time theory, constitutes the foundation also of the relativistic theory of gravitation. Only this theory can reveal the physical structure into which space-time order relations can be embedded even when all of the metrical properties of the space-time continuum are destroyed by gravitational fields. While we thus see in the causal theory of space and time the philosophical result of the theory of relativity, we wish to point out that this idea of a causal space-time order was conceived long before the advent of the theory of relativity. It was none other than Leibniz who developed in his "Initia rerum mathematicorum metaphysica" the basic ideas of this conception.[1] The new development is not a direct continuation of the work of Leibniz, who naturally did not know anything about the relativity of simultaneity. The author's development [2] of the causal theory of space and time was undertaken without knowledge of the corresponding work of Leibniz. It is the more remarkable that Leibniz, this genuine philosopher, was able to understand the nature of scientific knowledge to such an extent that, two hundred years later, a new development of physics and an analysis of its philosophic foundations confirmed his views.

C. The Most General Properties of Space and Time

§ 43. THE SINGULAR NATURE OF TIME

We shall now proceed to develop more precisely certain conclusions derivable from the causal theory of space and time that express very general properties of spatio-temporal order. A mathematical formulation will again provide us with a convenient approach to this problem. The fundamental topological properties of the special theory of relativity are expressed in the fundamental form of the metric, namely in its indefinite character and in its four-dimensionality. We shall discuss both of these factors in our presentation, devoting the present section to the indefinite character of the metric.

The indefinite character of the fundamental metrical form expresses the singular nature of time. We have pointed out, however, in § 29 that this characterization does not give us an exhaustive description of the nature of time. It expresses a distinction between space and time, but does not give us a complete comprehension of the peculiar

[1] See also H. Reichenbach, "Die Bewegungslehre bei Newton, Leibniz und Huyghens," *Kantstudien* 29, 1924, p. 421f.

[2] *Physikal. Zeitschr.* 22, 1921, p. 683; and in more detail in A. Similar ideas were developed by K. Lewin, *Ztschr. f. Phys.* 13, 62, 1923 and by R. Carnap, *Kantstudien* 30, 1925, p. 331.

characteristics of time. An additional essential property of time is its *directionality*. This specific feature is based on the fact that time—and time alone—is the dimension of the *causal chains* upon which we have based our theory of space and time. Time is the direction of the grain of the manifold along which the causal chains extend, whereas space reflects only the neighborhood relations between the coexisting causal chains. The direction of the causal chains is also the direction of the world-lines of objects which remain identical with themselves and which therefore represent special cases of causal chains. These lines exhibit most clearly the singular character of time. The atoms of a material rod located next to each other on a spacelike world-line are said to be different from one another; they are, of course, linked by dynamic bonds, but these forces merely combine them into a complex without destroying their individuality. The points of a time-like world-line, in contrast, are referred to as states of the *same* object. The atom of yesterday and the atom of today are *identical*, whereas the atom at the left end of the rod is *different* from the atom at the right end. We may denote this kind of identity by the term *genidentity*, introduced by K. Lewin.[1] Let us take, for example, a structure as complicated as the human organism. Mr. A of yesterday and Mr. A of today are identical, but not Mr. A and Mr. B. If this decisive difference did not exist between space and time, we could consider the continuation of yesterday's Mr. A to be today's (or even yesterday's) Mr. B, and we could construct the world-line of a human being running through several different individuals.

The theory of genidentity has received quite a blow from the criticism of the concept of substance, which was brought about by a reconsideration of the ether theory. Accordingly it is no longer necessary to consider the world-lines of a material field as striated in a definite direction; the choice of the grain includes a certain amount of arbitrariness. If the state of a field is graphically represented in the customary fashion, as in Fig. 46, then the vertical lines as well as the dotted slanted lines may be considered as the world-lines of the individual "field particles." Particle A_1 may thus be considered as genidentical with A_2, A_3 . . . as well as with B_2, C_3, D_4. . . . Nature does not supply a unique rule in this case. Einstein saw in this fact the collapse of the old concept of substance.[2] This means only (and

[1] K. Lewin, *Der Begriff der Genese*, Berlin 1922.

[2] A. Einstein: *Sidelights on Relativity*: I. *Ether and Relativity*. II. *Geometry and Experience*. London, 1922.

that is how it is formulated by Einstein) that *there are* material fields in which this arbitrariness exists. Einstein thus wishes to characterize the metrical field that propagates gravitational forces. On the other hand, there are also material fields in which there is a natural

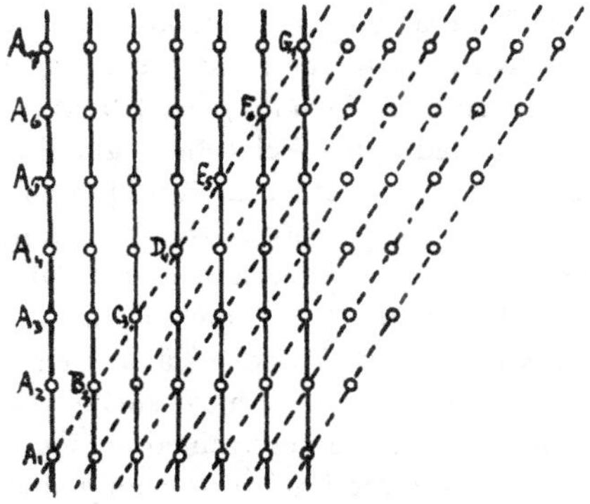

Fig. 46. Arbitrariness of the striation of world-lines in a continuous field.

striation; an example is atomic matter, the world-line bundles of which can by no means be considered as arbitrary in the sense of Fig. 46. We shall not pursue the significance of the concept of genidentity at this point, since this question would lead us away from our immediate problems of space and time into the problems of existence, which need not be referred to in this investigation because time preserves those properties with which we are concerned even in continuous fields, according to Fig. 46. The world-lines of Fig. 46 can be chosen arbitrarily only within the timelike cone, i.e., they cannot exceed a certain slope. Even continuous fields single out the dimension of time as uniquely suited to the concept of genidentity, a concept that can never be satisfied by spacelike world-lines.

The concept of genidentity is, consequently, closely related to the concept of causality. Different states can be genidentical only if they are causally related. This conception agrees with our definition of causal connection, which considers the causal chain a signal, i.e., the transmission of a mark. To speak of a recognition of the *same* mark implies a striation of the space-time manifold. Not all world-lines can be interpreted as lines of the progress of a mark. At the same time, however, our characterization of causality by means of the mark principle is sufficiently broad to include the arbitrariness in the choice of the lines of genidentity diagramed in Fig. 46.

If we now ask how the topological properties of time are to be characterized in more detail, we may point to § 21 and § 22 which contain the answer to this question. The laws formulated there as axioms of time order and time comparison contain those properties of time order that survive even in the most general gravitational fields. The fundamental concepts *earlier, later, indeterminate as to time order* are therefore left unaffected by the relativistic analysis of the concepts of space and time. These fundamental concepts constitute the core of the physical order of time. We notice that the *intuitive basis* of time order has been retained. The relativistic theory of gravitation does not destroy the intuitive character of time.

As explained, the validity of the topological axioms of time order can be asserted in gravitational fields only for *cut-out regions*. The structure of the world as a whole cannot be determined so long as the validity of special relativity in the infinitesimal is the only mathematical basis assumed. The holistic properties are not determined, if the description is restricted to infinitesimal regions; the topological character of the universe remains an open question. In particular the possibility remains that for the universe as a whole the axiom of the nonexistence of closed causal chains (p. 139) may fail. Mathematically speaking, it is possible to conceive a world in which the special theory of relativity applies in infinitesimal domains without exception, yet in which causal chains may be closed in the world as a whole.

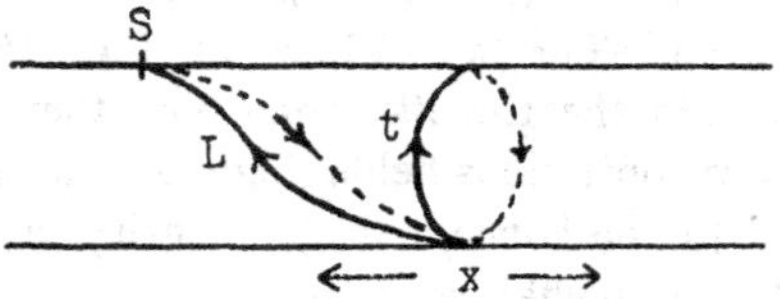

Fig. 47. A two-dimensional world, without singularities, containing closed time-lines.

To understand this result let us consider a graphical representation of a two-dimensional world, with one space and one time coordinate, drawn on the surface of a cylinder (Fig. 47). The time coordinate corresponds to a line encircling the cylinder, while the space coordinate corresponds to a line going to infinity in both directions, parallel to the axis of the cylinder. Such a manifold has no curvature and satisfies the special theory of relativity in every respect in finite cut-out regions. Only as a whole it has the peculiarity that time lines may be closed. A light signal L will travel like a spiral to infinity, yet it may be reflected (at S) so as to return to its starting point.

This simple model demonstrates strictly that closed timelike world-lines can exist in a world in which the special theory of relativity holds without singularities. Our proof is based on the idea that the Minkowskian geometrical representation of the indefinite metric accomplishes a one-to-one coordination between the definite and the indefinite metric and that a surface possible in a definite space will correspondingly lead to a coordinated surface in the indefinite space. This proof answers one of the questions posed in A., § 48. The proof of the satisfaction of axiom I, 1 in real systems (A., p. 148) is thus shown to be invalid by this model. The proof overlooks the fact that the existence of closed world-lines demands only an "apparent singularity" (A., p. 142) of the coordinate system, such as the singularity of the coordinate system of Fig. 47 at the origin of the time axis.

We described earlier (p. 141) the peculiar experiences that would result from the existence of closed causal chains. The general theory of relativity envisages such experiences as possible provided that no law regarding the order of space and time is recognized other than the special theory of relativity in infinitesimal domains. It may be advisable to consider this law insufficient and to exclude topological structures like that of Fig. 47 by a special axiom. At present, the evidence indicates that the time order is not closed.

These results exhaust the contribution of the theory of relativity to the problem of time. The theory of relativity has demonstrated the connection between the concepts of time and of causality and has led to the formulation of the general axioms of the order of time. A further penetration into the nature of time will be possible only if we subject the concept of causality to a more detailed analysis. This topic will have to be discussed in another publication.[1]

§ 44. THE NUMBER OF DIMENSIONS OF SPACE

Let us now turn to the second of the above-mentioned basic properties of the fundamental metrical form. Its indefinite character is expressed by the uniqueness of time. Now we shall investigate the significance of the number of its dimensions. Since one dimension is distinguished

[1] [See H. Reichenbach, *The Direction of Time*, University of California Press, Berkeley, 1956—M.R.]

from the others by virtue of the uniqueness of time, we can restrict our investigation to the three dimensions of space.

Various views have been expressed with regard to this question. It has been answered mostly in a subjective sense. The three-dimensionality of space has often been looked upon as a function of the human perceptual apparatus, which can visualize spatial relations only in this fashion. Poincaré tried to find a physiological foundation for this number; according to him two of the dimensions are due to the retinal image, the third to the "effort of accommodation which must be made, and to a sense of the convergence of the two eyes." [1] Even if this physiological explanation were tenable, it completely overlooks the fact that the number 3 of dimensions represents primarily a fact concerning the objective world and that the function of the visual apparatus is due to a developmental adaptation to the physical environment. This conception is supported by our account of *physical geometry*. In the present section we shall deal with the objective significance of the three-dimensionality of space.

We have previously shown (§ 12) that the topological properties of space are more directly determined by objective facts than its metrical properties. We showed that the redefinition of a geometry into a different topological type leads to causal anomalies. The same consequence applies to the number of dimensions, which is also a topological property. Two manifolds of different dimensionality can never be mapped upon one another in a continuous one-to-one transformation. The introduction of a different number of dimensions into physical space by means of a transformation—which is just as possible as the introduction of a different geometry—would destroy all existing causal laws. It is the characteristic of three-dimensionality that *it and only it* leads to continuous causal laws for physical reality. This empirical result describes a property of reality and constitutes the objective meaning of the statement that space has three dimensions.

We saw earlier that measurements of space are reducible to measurements of time and that in the most general gravitational fields, where no geometry of rigid bodies exists, the order of space can be defined only as the structure of causal chains. Now we can come to a deeper understanding of the connection between the dimensionality of space and the concept of causality. In § 27 we formulated definition 10 according to which a spatial distance is measured by the time a light

[1] *Science and Hypothesis*, Dover Publications, Inc., New York 1952, p. 53. See also p. 68.

ray requires to traverse it. If we now ignore the measure supplied by this definition, it will reduce to the more precise form of *definition e* (p. 169) of the concept *between*, whose properties are determined especially by axiom G. It has the same meaning as the *principle of action by contact*: causal effects cannot reach distant points of space without having previously passed through the intermediate points. Through the above definition of *between* the principle of action by contact becomes the more fundamental principle of spatial order: the neighborhood relations of space are to be chosen in such a way that the principle of action by contact is satisfied. This principle expresses the prescription which our concept of causality yields for the topology of space.

This rule determines the dimensionality of space as will be shown by the following consideration. First, it must be taken as an empirical fact that there is at least one dimensionality satisfying the principle of action by contact. If, however, there is such a dimensionality, there can be only *one*. Transformations to other dimensionalities always violate the principle of action by contact, because there are no continuous one-to-one transformations between spaces of different dimensionalities. This result is the basis for the determination of the number of dimensions. Let us now pursue this idea in more detail by means of a very simple example.

There are instances in physics where we work with spaces of a higher dimensionality, namely, whenever we use a so-called *parameter space*. Let us think, for example, of the state of a cloud of molecules as it occurs in a gas. The state of the gas is determined at any time t when the three coordinates of each of the n molecules are known, i.e., by the specification of n points in the three-dimensional *coordinate space*. Instead, we may consider all of the coordinates as dimensions of a $3n$-dimensional space, which is called the *parameter space*. The state of the gas is then given by *one* point in this $3n$-dimensional space. These descriptions are evidently equivalent: either of them can always be translated into the other. In spite of this fact we consider the parameter space merely a mathematical tool with no objective reference, whereas we regard the three-dimensional space as the *real space*. What is the justification for this distinction, which is ordinarily accepted as self-evident without further explanation? Is it not true that equivalent descriptions correspond equally to reality?

The answer is that these descriptions are not equivalent descriptions. The principle of action by contact decides in favor of the description in terms of the coordinate space. We shall explain this idea by the use of an example in which the parameter space has such a low number of dimensions that it can be diagramed in the customary fashion (Fig. 48). Let us consider two independent mass points moving on a straight line. In this case the coordinate space is one-dimensional and the parameter space is two-dimensional. The state of the point system at a given time is determined by the two coordinates x_1 and x_2 of the mass points.

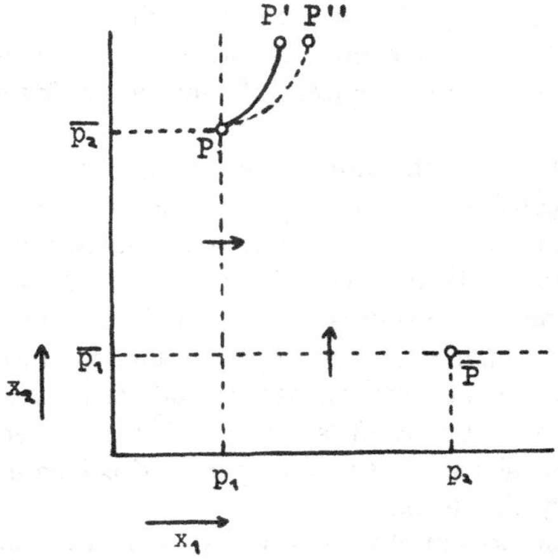

Fig. 48. Propagation of a disturbance in the parameter space.

Hence, the location of the points is given *either* by the two points p_1 and p_2 in the 1-dimensional coordinate space *or* by a single point P in the parameter space. Formally speaking, we may conceive the motion of the point system either as the motion of two points in a 1-dimensional space or as the motion of a single point in a 2-dimensional space. But if we now introduce the principle of action by contact, the descriptions cease to be equivalent.

Let us consider a disturbance in the 1-dimensional space, such as a sound wave originating from point p_1. According to the principle of action by contact, it will gradually travel from p_1 in both directions. What would be its configuration in the 2-dimensional parameter space? Every point P of the plane corresponds to a combination of two points $p_i p_k$ on the x_1 axis. All combinations $p_1 p_k$ lie on the straight line $p_1 P$. Since the disturbance acts at p_1, it must therefore affect all points on the line p_1P simultaneously and the propagation of the disturbance will be represented by a lateral shift of the line p_1P in the direction of the arrow. At the same time, however, there must also occur a horizontal, symmetrically located, disturbance front $\bar{p}_1\bar{P}$, since p_2 will also be disturbed as soon as it approaches the area of the disturbance. Whereas the disturbance spreads from a center in the coordinate space, it does not have a central point in the parameter space, but affects immediately two intersecting straight lines. This fact shows that the "real space" in this example is 1-dimensional; only in the 1-dimensional space can the principle of action by contact be maintained, while in the 2-dimensional space infinite velocities of propagation exist along the intersecting straight lines.

Besides this interpretation, there exists another in which we do not speak of infinite speeds of propagation along the straight line, but of a preestablished harmony. Whenever a disturbance originates in a point on one of the straight lines, it originates at the same time also in all of its other points, because there is a corresponding cause for the disturbance at every point. These two interpretations are equivalent and constitute the same violation of the principle of causality. The satisfaction of the principle of action by contact excludes, therefore, preestablished harmony. See also p. 65.

We must now formulate this analysis in more detail. The transition from the coordinate space to the parameter space involves a certain arbitrariness, and we must now specify the presupposition according to which the point disturbance in the coordinate space is transformed into a line disturbance in the parameter space. If p_1 lies within the region of the disturbance, its path will be deflected. Consequently, the path of the combination point P will also be deflected. and it is irrelevant in this case whether p_2 is affected by the disturbance. If we now require that a deflection of P can occur only if P lies within the domain of disturbance in the parameter space, it follows necessarily that the disturbance in the parameter space must occur along a line and cannot be restricted to the immediate neighborhood of a point.

This disturbance has a second peculiarity. At first it has no effect on the state of motion of p_2 since it affects only p_1. In the 2-dimensional parameter space, it can therefore affect the point P of the system only in a certain limited fashion, namely, the disturbance can change only the x_1 coordinate and not the x_2 coordinate of P. If PP' is a segment of the undisturbed path of P, and PP'' is the corresponding segment of the disturbed path, then the points P' and P'' lying on the same horizontal line must correspond to simultaneous events. The path of P receives from the linear disturbance only a sideward bulge perpendicular to the front of the disturbance. The central disturbance spreading in all directions in the coordinate space will be linear and will spread unidirectionally in the parameter space.

If, on the other hand, the process of the disturbance were centrally oriented in the 2-dimensional space, it would not be oriented in the same way in the 1-dimensional space. If we consider a disturbance spreading centrally from P, it will be represented in the 1-dimensional space by two disturbances which affect p_1 and p_2 simultaneously and spread centrally from each of these points. Between these two disturbances there would exist an action at a distance: if the point p_1 should enter one of the regions of disturbance, it would be disturbed only if simultaneously the distant point p_2 were located in the other region of disturbance.[1] A central action by contact in the parameter space would therefore lead to an action at a distance in the coordinate space. In this case we would say that the 2-dimensional space is the "real one," while the 1-dimensional space is merely a mathematical tool.

As was pointed out before, the described difference in the causal propagation of the two spaces is based on the fact that no continuous one-to-one transformation is possible between spaces of different dimensionalities. There can only be a transformation that changes the elements of the space, i.e., a point in the 2-dimensional space corresponds to a combination of points in the 1-dimensional space, and a point in the 1-dimensional space corresponds to a straight line (or a pair of intersecting straight lines) in the 2-dimensional space.

An extension of these considerations to higher numbers of dimensions is easily accomplished. If the coordinate space is 3-dimensional, and the parameter space n-dimensional ($n > 3$), then a disturbance in the coordinate space which obeys the principle of action by contact is represented in the parameter space by intersecting hyperplanes moving sideward, which can again be disturbed only in a limited fashion and which will cover, in the course of time, a cylindrical region having a 3-dimensional cross-section. A centro-symmetric disturbance in the coordinate space corresponds therefore to a group of cylindrical and unidirectional disturbances in the parameter space. Conversely, a centro-symmetric disturbance in the parameter space would be

[1] This follows because the combination point P lies in the region of the disturbance of the parameter space only under these conditions. The above-mentioned requirement that P is disturbed only under these conditions is here obvious, since otherwise we cannot speak of action by contact in the parameter space.

equivalent in the coordinate space to a number of separate disturbances connected at a distance.

Let us now imagine schematically that an observer wants to determine experimentally the dimensionality of space. He will try to combine the parameters of the observed events into one space and test whether the principle of action by contact is satisfied in this manifold. If it is not, he will try a different combination of parameters, i.e., a different dimensionality of the parameter space, testing thus various parameter spaces until he finds one that has the desired property. This space he will call the coordinate space. If we consider the coordinate space as a special case of a parameter space, we may formulate the results of our considerations as follows:

The principle of action by contact can be satisfied only for a single choice of the dimensionality of the parameter space; that particular parameter space in which it is satisfied is called the coordinate space or "real space."

This formulation expresses the requirements on which the coordinative definition of the topological space is based. The topology is therefore basically subject to the same qualifications as the metric: without a coordinative definition it is not determined, and therefore we cannot regard it as an absolute datum. The metric of a space becomes an empirical fact only after the postulate of the disappearance of universal forces is introduced. Similarly, the topology of space becomes an empirical fact only if we add the postulate of the principle of action by contact. This idea was first considered in § 12, where we dealt with the topological character of space and recognized that it is determined only if we add the postulate of the disappearance of causal anomalies. This result fits into our present more general requirement, according to which not only the topological character but also the neighborhood relations are determined only if the postulate of action by contact is assumed. A recognition of the arbitrary components in the various descriptions clarifies the objective nature of topology. That the requirement of the principle of action by contact can be satisfied at all, and in particular, that it is satisfied in a Riemannian space of three dimensions, is a physical fact not dependent on arbitrary definitions. *The statement that physical space has three dimensions has therefore the same objective character as, for instance, the statement that there are three physical states of matter, the solid, liquid, and gaseous state; it describes a fundamental fact of the objective world.*

Just as in the case of the metric, one can proceed, after recognizing that the three-dimensionality is a physical fact, to the question of its explanation, i.e., one can now search for a cause of the three-dimensionality of space. One might try, for instance, to regard the three-dimensionality as a consequence of certain conditions of equilibrium of matter. This conception would be justified if the three-dimensional order of matter could be shown to be the only stable order. Any such proof presupposes certain laws of nature which can be formulated *independently of the dimensionality of space.* Such a proof might read: If space has n dimensions, and it is a general law of nature that the attraction between masses varies inversely with the (n-1)th power of their distance, then the dimensionality of space must be $n = 3$, since otherwise the motion of the planets and also the arrangement of the masses of the stars would not be stable. Of course, this idea cannot be carried over into Einsteinian geometry, where the Newtonian force of attraction no longer plays a primary role and occurs only as an approximate solution. The proof could also be based on a more general assumption, such as the following: If the field equations of matter contain only differentials of second order or less, space must have three dimensions. We mention these examples, which of course are not intended to represent confirmed laws of nature, nor laws for whose future confirmation we see any chance, merely in order to express the basic idea of such an explanation of the dimensionality of space. This explanation, like any other explanation, can consist only in a combination of two natural phenomena into one, i.e., in the derivation of one from the other. The three-dimensionality would thus be recognized as a logical consequence of certain fundamental properties of matter, which in turn would have to be accepted as ultimate facts. Any other attempt at explanation would be vain. The three-dimensionality of space cannot be maintained as an absolute necessity; it is a physical fact like any other, and therefore subject to the same kind of explanation. Though some attempts have been made to treat the problem from this point of view, as for instance by H. Weyl[1] and P. Ehrenfest,[2] they have thus far not led to success.

We shall now turn briefly to the question of the visualization of spaces of higher dimensionality. There can be no doubt that in a world of a higher dimensionality the human power of imagination would adapt itself to its environment and that man would have a visual picture of this space analogous to his present three-dimensional visualizations. When we try to imagine such a picture according to the rule given by Helmholtz (p. 63), namely, by describing experiences in a four-

[1] *Op. cit.*, p. 285.

[2] *Ann. d. Phys.* 61, 1920, p. 440.

dimensional space (i.e., a five-dimensional space-time manifold), we meet with certain difficulties. In such a space even the human body would be four-dimensional, and its perceptual apparatus would be very different. Instead of the two-dimensional retina of the eye, there would be a three-dimensional retina. Whereas the visual experience of the third dimension, the "depth," is now achieved primarily by the combined effect of the two eyes and is therefore qualitatively different from the experience of the other two dimensions, the three-dimensional experiences in a four-dimensional space would be as immediate as two-dimensional experiences in our three-dimensional space. The combined effect of the two three-dimensional pictures on the retina would supply us with the visual experience of the four-dimensional space. If we try to imagine such experiences in terms of our present sensations, we shall find that there are certain limitations. The new perceptual experience we wish to describe would have new sense qualities that do not exist under the conditions with which we are familiar.

We can therefore indicate only indirectly what kind of experiences would result in such a world. For this purpose we shall use the device of substituting for the new, unknown sense quality a known quality. Let us assume that the three dimensions of space are visualized in the customary fashion, and let us substitute a color for the fourth dimension. Every physical object is liable to changes in color as well as in position. An object might, for example, be capable of going through all shades from red through violet to blue. A physical interaction between any two bodies is possible only if they are close to each other in space as well as in color. Bodies of different colors would penetrate each other without interference. In this fashion we have now coordinated to every point of the three-dimensional space the one-dimensional continuous manifold of color from red through violet to blue. Hence the combined manifold of these states is four-dimensional. This manifold constitutes a space in the proper sense because the principle of action by contact is satisfied.

All properties of a four-dimensional space may now be inferred from this illustration. The collision of two billiard-balls, for instance, would occur as follows: the two balls approach the same three-dimensional point, while at the same time their colors become more and more alike. Only if they are close not only in space, but also in color, will there be the sound of a collision. It may also happen that the two balls are at the same point in space; but so long as their colors are different, they will penetrate each other. If they stay at the same spot and their colors become gradually more alike, we shall finally hear the sound of their collision at the moment when their colors are identical. The fact that a closed three-dimensional surface no longer encloses a spatial region will become clear from the following consideration. If we lock a number of flies into a red glass globe, they may yet escape: they may change their color to blue and are then able to penetrate the red globe.

The human body would be differently built in such a world. It would extend not only in the three dimensions of space but also in the dimension of color. It would consist of similar bodies which penetrate one another and are somewhat different in respect to color, thus filling in continuous variation a small interval of color. The retina would also consist of such interpenetrating layers. Any bundle of light rays striking the eye would have an additional dimension. If it were a two-dimensional bundle in the three-dimensional space, it would be three-dimensional in the four-dimensional space. We must here imagine that every light ray is capable of assuming every color. From a red point in space, a red ray will go to a red point of the retina. The same point of the retina is "penetrated" completely by a blue ray traveling from a corresponding blue point in space to the corresponding blue point of the retina. The new sense quality can now be indicated as follows. Since the corresponding red and blue rays are received by different elements of the retina, we would perceive them as differently located in space, just as we perceive the direction of rays as spatially different in three dimensions if they hit different elements of the retina. This difference in spatial localization, which was replaced in our example by a difference in color, cannot be perceived by three-dimensional human beings. The four-dimensional human being would perceive the gradually increasing similarity of the colors of the billiard balls as a change in spatial position, just as we experience the decreasing distance between the balls in three dimensions as a change in their positions due to the successive stimulation of different cells of the retina. These four-dimensional changes of position, of course, need not be connected with the phenomenon of color, since we have used color merely as a device.

Here we encounter a limit to our capacity for visualization. New sense qualities cannot be predicted: we can only use substitutes for them. This difficulty expresses the fundamental importance of dimensionality. The visualization of events that take place in a metrically different space, or in a topologically different sp; dimensions, is possible because these spaces are Euclid infinitesimal, and perceptual experiences are therefore essentially unchanged. An increase in the number of dimensions, however, affects even the smallest regions and provides therefore a qualitatively different experience.

We can conceive special cases in which the space is four-dimensional, yet in which the perceptual experience is not different from that in three-dimensional space. Such a case would occur if the human body, as a three-dimensional structure, were embedded in a four-dimensional space. In this case we can imagine what the three-dimensionally constructed human eye would see in the four-dimensional space. This situation corresponds to one in which two-dimensional human beings live in a three-dimensional space and are able to perceive it. However, since there seems to be a physical law that objects capable of physical existence must have as many dimensions as the surrounding space, this example corresponds to a world which cannot be physically realized. Another case would arise if space were four- (or more) dimensional in its smallest elements, but three-dimensional as a whole. This situation would correspond to the case of a thin layer of grains of sand which, although each is three-dimensional if taken individually, taken as a whole forms essentially a two-dimensional space. Similarly, atoms which individually are higher-dimensional could cluster into three-dimensional structures. In such a world, a macroscopic structure would have only the degrees of freedom of the three dimensions of space, while an atom would have many more degrees of freedom. Sense perceptions in such a world would not be noticeably different from those of our ordinary world; and conversely, it is in principle possible to infer from our ordinary experiences the higher-dimensional character of the microscopic world. Incidentally, it is not impossible that quantum mechanics will lead to such results.

§ 45. THE REALITY OF SPACE AND TIME

Statements about the topological properties turn out to be the most reliable ones we can make about the order of space and time; they apply even to the most general gravitational fields. We said in § 39 that the Gauss-Riemann separation of coordinate system and metric leaves to the coordinate system the function of characterizing the topological properties of a space. We may therefore regard the following statement as the most general assertion about space-time order: everywhere and at all times there exists a *space-time coordinate system*.

This result implies the *topological distinguishability of space and time*. In a space-time coordinate system one of the dimensions is to be considered as time and the three others as space. The division into timelike and spacelike directions is accomplished by the world-lines of light. In Fig. 49 the dividing lines are drawn, in Fig. 49*a* for the special theory of relativity, in Fig. 49*b* for the general theory of relativity.

The second diagram differs considerably from the first in its metric, while it is topologically equivalent to it. In both diagrams we have indicated by a dotted line a timelike world-line (with arrow) and a spacelike world-line (without arrow). It can easily be seen how the angular regions traversed by these lines can be characterized topologically with reference to the arrows of the world-lines of light: timelike world-lines pass through the angular region between those branches of the lines that have arrows, while spacelike world-lines pass through the angular region between one branch with an arrow and one without an arrow. If we were to draw in the same manner a set of timelike and a set of spacelike world-lines, they would represent a space-time coordinate system (a real system).

It is of course not necessary to describe a manifold of the type of Fig. 49 by such a coordinate system. We could obtain a unique description also by means of two sets of spacelike lines. However, the fact that a topologically divided coordinate system *can be chosen* expresses a most important property of the physical world.

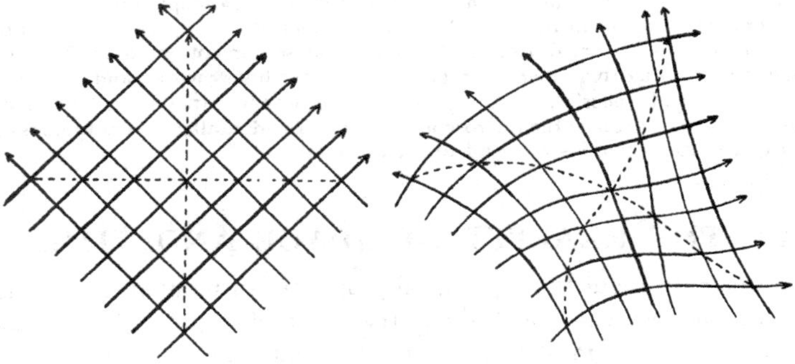

a b

Fig. 49 *a* and *b*. The topological divisibility of a space-time manifold.

We can visualize the situation regarding the space-time manifold as follows. A space-time coordinate system is obtained if space points are numbered in some continuous fashion and time is measured by clocks the simultaneity of which is defined in such a manner that it connects only events indeterminate as to time order (as described in A., p. 152). But we could replace the time determination by a completely different one. Let us suppose that all over the space there is a downpour of projectiles which are numbered consecutively. Instead of the time indication we may then use as the fourth coordinate the number of the projectile closest to the event to be described. The coordinate system thus obtained is not a real system (its simultaneity cross-section lies in the region of temporal succession), but it supplies a unique determination of all of the events. We dealt with an unreal system of a different kind in § 38, where we recognized that a rotating system of axes becomes unreal beyond a certain distance from the center. We therefore cannot say that the choice of a space-time divided coordinate system is *necessary*; but that it is always *possible* means the same as that the space-time manifold can be divided topologically.

This result leads to the question whether the possibility of such a choice of coordinates is derivable from the validity of the special theory of relativity in infinitesimal domains. For the time being such a proof is possible only for cut-out regions. See also A., p. 154. The possibility of such a choice of coordinates for the world as a whole would therefore have to be asserted as a special axiom.

The fact that an ordering of all events is possible within the three dimensions of space and the one dimension of time is the most fundamental aspect of the physical theory of space and time. In comparison, the possibility of a metric seems to be of subordinate significance. It is only the metric, however, which, in the general theory of relativity has been recognized as an effect of the gravitational field. The essence of space-time order, its topology, remains an ultimate fact of nature, unaffected by these considerations. We must therefore content ourselves with the statement that an ordering in terms of space-time coordinates is possible. The explanation of topology would still be incomplete even if we were to find the explanation of three-dimensionality discussed in the previous section. The three-dimensionality represents one of the topological properties of space and time, and any explanation would have to start with the assumption that some continuous order of space and time exists.

Other attempts have been made to explain the topology of space and time. The coordinate system assigns to the system of coincidences, of point-events, a mutual order that is independent of any metric. This order of coincidences must therefore be understood as an ultimate fact. The attempt has been made to justify this order as necessary; it has been regarded as a function of the human perceptual apparatus rather than of the objective world. Accordingly, it has been claimed that sense perceptions supply directly only coincidences, and that the ultimate element of space-time order is determined by the character of our sense perceptions. In this connection appeal is made to the experimental methods of the physicist, in which coincidences of dials and pointers play an important role.

This view is untenable. First of all, it is obvious that we cannot regard the order of coincidences as immediately given, since the subjective order of perceptions does not necessarily correspond to the objective order of external events. It can serve only as the basis of a complicated procedure by which the objective order is inferred. This difference is due to the fact that perceptions form a one-dimensional chain, while objective point events belong to a four-dimensional manifold. We must therefore introduce rules for the construction of the objective order; such rules have been formulated in the topological coordinative definitions.

It is a serious mistake to identify a coincidence, in the sense of a point-event of the space-time order, with a coincidence in the sense of a sense experience.[1] The latter is *subjective coincidence,* in which sense perceptions are blended; for instance, the experience of sound can be blended with the impression of light. The former, on the other hand, is *objective* coincidence, in which physical things, such as atoms, billiard balls or light rays collide and which can take place even when no observer is present. The space-time order deals only with objective coincidences, and we go outside the realm of its problems in asking how the system of objective coincidences is related to the corresponding subjective system. The analysis of this question belongs to that part of epistemology that explains the connection between objective reality, on the one hand, and consciousness and perception on the other. Let us say here only that any statement about objective coincidences has the same epistemological status as any other statement concerning a physical fact.

It is therefore not possible to reduce the topology of space and time to subjective grounds springing from the nature of the observer. On the contrary, we must specify the principles according to which an objective coincidence is to be ascertained. This means that we must indicate a method how to decide whether a physical event is to be considered as one, or as two or more separate point events. Such methods are frequently employed by the physicist, although he is usually not aware of this fact. He decides, for example, that the movement of a particle in Brownian motion is to be represented not as

[1] See also A., § 4, for this distinction.

one coincidence, but as a great number of spatially and temporally separate coincidences, whose integral effect is observed. The general basis of all such procedures was explained in § 43, where we demonstrated that the principle of action by contact is decisive for the determination of the dimensionality of space. In the example given there, we investigated the question whether a certain event is to be considered as one point-event in a two-dimensional space, or as two point-events in a one-dimensional space. Any such decision depends on the question, in which coordinate system the principle of action by contact is satisfied, although both systems appear at first to be equally adequate. This procedure is used by the physicist to decide what constitutes a point-event: occurrences are point-events, if the assumption that they are point-events, in combination with observation, leads to the conclusion that the principle of action by contact is satisfied.

Objective coincidences are therefore physical events like any others; their occurrence can be confirmed only within the context of theoretical investigation. Since all happenings have until now been reducible to objective coincidences, we must consider it the most general empirical fact that the physical world is a system of coincidences. It is this fact on which all spatio-temporal order is based, even in the most complicated gravitational fields. What kind of physical occurrences are coincidences, however, is not uniquely determined by empirical evidence, but depends again on the totality of our theoretical knowledge.

The most important result of these considerations is the objectivity of the properties of space. *The reality of space and time* turns out to be the irrefutable consequence of our epistemological analyses, which have led us through many important individual problems. This result is somewhat obscured by the appearance of an element of arbitrariness in the choice of the description. But in showing that the arbitrariness pertains to coordinative definitions we could make a precise statement about the empirical component of all space-time descriptions. Philosophers have thus far considered an idealistic interpretation of space and time as the only possible epistemological position, because they overlooked the twofold nature of the mathematical and the physical problems of space. Mathematical space is a *conceptual structure*, and as such *ideal*. Physics has the task of coordinating one of these mathematical structures to *reality*. In fulfilling this task, physics makes statements about reality, and it has been our aim to free the objective core of these assertions from the subjective additions introduced through the arbitrariness in the choice of the description.

A deeper understanding of this problem would of course be possible only if we were to enter into a detailed analysis of the problem of reality and description, i.e., the problem of physical knowledge in general. It may suffice at this place to remark that the problem concerning space and time is not different from that of the description of any other physical state as expressed in physical laws.

If the properties of space and time reflect laws of nature, however, their physical treatment requires an essential correction. We can employ the concepts of space and time only so long as there are phenomena that realize them. At present the existence of such phenomena has been confirmed only for the macrocosm. All of the physical concepts basic to the order of space and time, namely, rigid bodies, clocks, causal chains, coincidences, refer to macrocosmic phenomena. We cannot be sure that they can be extended to the microcosm, i.e., the world of the interior of the atom. In addition to the concept of coincidence, only the concept of causal chain has some prospects for such an extension. If we have hopes that a space-time order can be constructed even for the smallest elements of matter, this optimism derives from the circumstance that it was the concept of causal chain which proved to be the ultimate basis of the space-time order of the macrocosm, when all other means to establish an order had failed in the most general gravitational fields. A final decision with regard to this problem must be postponed, however, until the problem of matter, at present in the focus of scientific research, has been solved.[1]

[1] [See H. Reichenbach, *The Direction of Time*, University of California Press, Berkeley, 1956, Ch. V.—M.R.]

INDEX

A CATALOG OF SELECTED

DOVER BOOKS

IN SCIENCE AND MATHEMATICS

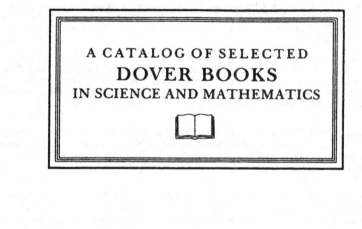

Astronomy

CHARIOTS FOR APOLLO: The NASA History of Manned Lunar Spacecraft to 1969, Courtney G. Brooks, James M. Grimwood, and Loyd S. Swenson, Jr. This illustrated history by a trio of experts is the definitive reference on the Apollo spacecraft and lunar modules. It traces the vehicles' design, development, and operation in space. More than 100 photographs and illustrations. 576pp. 6 3/4 x 9 1/4. 0-486-46756-2

EXPLORING THE MOON THROUGH BINOCULARS AND SMALL TELESCOPES, Ernest H. Cherrington, Jr. Informative, profusely illustrated guide to locating and identifying craters, rills, seas, mountains, other lunar features. Newly revised and updated with special section of new photos. Over 100 photos and diagrams. 240pp. 8 1/4 x 11. 0-486-24491-1

WHERE NO MAN HAS GONE BEFORE: A History of NASA's Apollo Lunar Expeditions, William David Compton. Introduction by Paul Dickson. This official NASA history traces behind-the-scenes conflicts and cooperation between scientists and engineers. The first half concerns preparations for the Moon landings, and the second half documents the flights that followed Apollo 11. 1989 edition. 432pp. 7 x 10. 0-486-47888-2

APOLLO EXPEDITIONS TO THE MOON: The NASA History, Edited by Edgar M. Cortright. Official NASA publication marks the 40th anniversary of the first lunar landing and features essays by project participants recalling engineering and administrative challenges. Accessible, jargon-free accounts, highlighted by numerous illustrations. 336pp. 8 3/8 x 10 7/8. 0-486-47175-6

ON MARS: Exploration of the Red Planet, 1958-1978--The NASA History, Edward Clinton Ezell and Linda Neuman Ezell. NASA's official history chronicles the start of our explorations of our planetary neighbor. It recounts cooperation among government, industry, and academia, and it features dozens of photos from Viking cameras. 560pp. 6 3/4 x 9 1/4. 0-486-46757-0

ARISTARCHUS OF SAMOS: The Ancient Copernicus, Sir Thomas Heath. Heath's history of astronomy ranges from Homer and Hesiod to Aristarchus and includes quotes from numerous thinkers, compilers, and scholasticists from Thales and Anaximander through Pythagoras, Plato, Aristotle, and Heraclides. 34 figures. 448pp. 5 3/8 x 8 1/2. 0-486-43886-4

AN INTRODUCTION TO CELESTIAL MECHANICS, Forest Ray Moulton. Classic text still unsurpassed in presentation of fundamental principles. Covers rectilinear motion, central forces, problems of two and three bodies, much more. Includes over 200 problems, some with answers. 437pp. 5 3/8 x 8 1/2. 0-486-64687-4

BEYOND THE ATMOSPHERE: Early Years of Space Science, Homer E. Newell. This exciting survey is the work of a top NASA administrator who chronicles technological advances, the relationship of space science to general science, and the space program's social, political, and economic contexts. 528pp. 6 3/4 x 9 1/4. 0-486-47464-X

STAR LORE: Myths, Legends, and Facts, William Tyler Olcott. Captivating retellings of the origins and histories of ancient star groups include Pegasus, Ursa Major, Pleiades, signs of the zodiac, and other constellations. "Classic." – *Sky & Telescope.* 58 illustrations. 544pp. 5 3/8 x 8 1/2. 0-486-43581-4

A COMPLETE MANUAL OF AMATEUR ASTRONOMY: Tools and Techniques for Astronomical Observations, P. Clay Sherrod with Thomas L. Koed. Concise, highly readable book discusses the selection, set-up, and maintenance of a telescope; amateur studies of the sun; lunar topography and occultations; and more. 124 figures. 26 halftones. 37 tables. 335pp. 6 1/2 x 9 1/4. 0-486-42820-6

Chemistry

MOLECULAR COLLISION THEORY, M. S. Child. This high-level monograph offers an analytical treatment of classical scattering by a central force, quantum scattering by a central force, elastic scattering phase shifts, and semi-classical elastic scattering. 1974 edition. 310pp. 5 3/8 x 8 1/2. 0-486-69437-2

HANDBOOK OF COMPUTATIONAL QUANTUM CHEMISTRY, David B. Cook. This comprehensive text provides upper-level undergraduates and graduate students with an accessible introduction to the implementation of quantum ideas in molecular modeling, exploring practical applications alongside theoretical explanations. 1998 edition. 832pp. 5 3/8 x 8 1/2. 0-486-44307-8

RADIOACTIVE SUBSTANCES, Marie Curie. The celebrated scientist's thesis, which directly preceded her 1903 Nobel Prize, discusses establishing atomic character of radioactivity; extraction from pitchblende of polonium and radium; isolation of pure radium chloride; more. 96pp. 5 3/8 x 8 1/2. 0-486-42550-9

CHEMICAL MAGIC, Leonard A. Ford. Classic guide provides intriguing entertainment while elucidating sound scientific principles, with more than 100 unusual stunts: cold fire, dust explosions, a nylon rope trick, a disappearing beaker, much more. 128pp. 5 3/8 x 8 1/2. 0-486-67628-5

ALCHEMY, E. J. Holmyard. Classic study by noted authority covers 2,000 years of alchemical history: religious, mystical overtones; apparatus; signs, symbols, and secret terms; advent of scientific method, much more. Illustrated. 320pp. 5 3/8 x 8 1/2. 0-486-26298-7

CHEMICAL KINETICS AND REACTION DYNAMICS, Paul L. Houston. This text teaches the principles underlying modern chemical kinetics in a clear, direct fashion, using several examples to enhance basic understanding. Solutions to selected problems. 2001 edition. 352pp. 8 3/8 x 11. 0-486-45334-0

PROBLEMS AND SOLUTIONS IN QUANTUM CHEMISTRY AND PHYSICS, Charles S. Johnson and Lee G. Pedersen. Unusually varied problems, with detailed solutions, cover of quantum mechanics, wave mechanics, angular momentum, molecular spectroscopy, scattering theory, more. 280 problems, plus 139 supplementary exercises. 430pp. 6 1/2 x 9 1/4. 0-486-65236-X

ELEMENTS OF CHEMISTRY, Antoine Lavoisier. Monumental classic by the founder of modern chemistry features first explicit statement of law of conservation of matter in chemical change, and more. Facsimile reprint of original (1790) Kerr translation. 539pp. 5 3/8 x 8 1/2. 0-486-64624-6

MAGNETISM AND TRANSITION METAL COMPLEXES, F. E. Mabbs and D. J. Machin. A detailed view of the calculation methods involved in the magnetic properties of transition metal complexes, this volume offers sufficient background for original work in the field. 1973 edition. 240pp. 5 3/8 x 8 1/2. 0-486-46284-6

GENERAL CHEMISTRY, Linus Pauling. Revised third edition of classic first-year text by Nobel laureate. Atomic and molecular structure, quantum mechanics, statistical mechanics, thermodynamics correlated with descriptive chemistry. Problems. 992pp. 5 3/8 x 8 1/2. 0-486-65622-5

ELECTROLYTE SOLUTIONS: Second Revised Edition, R. A. Robinson and R. H. Stokes. Classic text deals primarily with measurement, interpretation of conductance, chemical potential, and diffusion in electrolyte solutions. Detailed theoretical interpretations, plus extensive tables of thermodynamic and transport properties. 1970 edition. 590pp. 5 3/8 x 8 1/2. 0-486-42225-9

Engineering

FUNDAMENTALS OF ASTRODYNAMICS, Roger R. Bate, Donald D. Mueller, and Jerry E. White. Teaching text developed by U.S. Air Force Academy develops the basic two-body and n-body equations of motion; orbit determination; classical orbital elements, coordinate transformations; differential correction; more. 1971 edition. 455pp. 5 3/8 x 8 1/2. 0-486-60061-0

INTRODUCTION TO CONTINUUM MECHANICS FOR ENGINEERS: Revised Edition, Ray M. Bowen. This self-contained text introduces classical continuum models within a modern framework. Its numerous exercises illustrate the governing principles, linearizations, and other approximations that constitute classical continuum models. 2007 edition. 320pp. 6 1/8 x 9 1/4. 0-486-47460-7

ENGINEERING MECHANICS FOR STRUCTURES, Louis L. Bucciarelli. This text explores the mechanics of solids and statics as well as the strength of materials and elasticity theory. Its many design exercises encourage creative initiative and systems thinking. 2009 edition. 320pp. 6 1/8 x 9 1/4. 0-486-46855-0

FEEDBACK CONTROL THEORY, John C. Doyle, Bruce A. Francis and Allen R. Tannenbaum. This excellent introduction to feedback control system design offers a theoretical approach that captures the essential issues and can be applied to a wide range of practical problems. 1992 edition. 224pp. 6 1/2 x 9 1/4. 0-486-46933-6

THE FORCES OF MATTER, Michael Faraday. These lectures by a famous inventor offer an easy-to-understand introduction to the interactions of the universe's physical forces. Six essays explore gravitation, cohesion, chemical affinity, heat, magnetism, and electricity. 1993 edition. 96pp. 5 3/8 x 8 1/2. 0-486-47482-8

DYNAMICS, Lawrence E. Goodman and William H. Warner. Beginning engineering text introduces calculus of vectors, particle motion, dynamics of particle systems and plane rigid bodies, technical applications in plane motions, and more. Exercises and answers in every chapter. 619pp. 5 3/8 x 8 1/2. 0-486-42006-X

ADAPTIVE FILTERING PREDICTION AND CONTROL, Graham C. Goodwin and Kwai Sang Sin. This unified survey focuses on linear discrete-time systems and explores natural extensions to nonlinear systems. It emphasizes discrete-time systems, summarizing theoretical and practical aspects of a large class of adaptive algorithms. 1984 edition. 560pp. 6 1/2 x 9 1/4. 0-486-46932-8

INDUCTANCE CALCULATIONS, Frederick W. Grover. This authoritative reference enables the design of virtually every type of inductor. It features a single simple formula for each type of inductor, together with tables containing essential numerical factors. 1946 edition. 304pp. 5 3/8 x 8 1/2. 0-486-47440-2

THERMODYNAMICS: Foundations and Applications, Elias P. Gyftopoulos and Gian Paolo Beretta. Designed by two MIT professors, this authoritative text discusses basic concepts and applications in detail, emphasizing generality, definitions, and logical consistency. More than 300 solved problems cover realistic energy systems and processes. 800pp. 6 1/8 x 9 1/4. 0-486-43932-1

THE FINITE ELEMENT METHOD: Linear Static and Dynamic Finite Element Analysis, Thomas J. R. Hughes. Text for students without in-depth mathematical training, this text includes a comprehensive presentation and analysis of algorithms of time-dependent phenomena plus beam, plate, and shell theories. Solution guide available upon request. 672pp. 6 1/2 x 9 1/4. 0-486-41181-8

Mathematics-Bestsellers

HANDBOOK OF MATHEMATICAL FUNCTIONS: with Formulas, Graphs, and Mathematical Tables, Edited by Milton Abramowitz and Irene A. Stegun. A classic resource for working with special functions, standard trig, and exponential logarithmic definitions and extensions, it features 29 sets of tables, some to as high as 20 places. 1046pp. 8 x 10 1/2. 0-486-61272-4

ABSTRACT AND CONCRETE CATEGORIES: The Joy of Cats, Jiri Adamek, Horst Herrlich, and George E. Strecker. This up-to-date introductory treatment employs category theory to explore the theory of structures. Its unique approach stresses concrete categories and presents a systematic view of factorization structures. Numerous examples. 1990 edition, updated 2004. 528pp. 6 1/8 x 9 1/4. 0-486-46934-4

MATHEMATICS: Its Content, Methods and Meaning, A. D. Aleksandrov, A. N. Kolmogorov, and M. A. Lavrent'ev. Major survey offers comprehensive, coherent discussions of analytic geometry, algebra, differential equations, calculus of variations, functions of a complex variable, prime numbers, linear and non-Euclidean geometry, topology, functional analysis, more. 1963 edition. 1120pp. 5 3/8 x 8 1/2. 0-486-40916-3

INTRODUCTION TO VECTORS AND TENSORS: Second Edition--Two Volumes Bound as One, Ray M. Bowen and C.-C. Wang. Convenient single-volume compilation of two texts offers both introduction and in-depth survey. Geared toward engineering and science students rather than mathematicians, it focuses on physics and engineering applications. 1976 edition. 560pp. 6 1/2 x 9 1/4. 0-486-46914-X

AN INTRODUCTION TO ORTHOGONAL POLYNOMIALS, Theodore S. Chihara. Concise introduction covers general elementary theory, including the representation theorem and distribution functions, continued fractions and chain sequences, the recurrence formula, special functions, and some specific systems. 1978 edition. 272pp. 5 3/8 x 8 1/2. 0-486-47929-3

ADVANCED MATHEMATICS FOR ENGINEERS AND SCIENTISTS, Paul DuChateau. This primary text and supplemental reference focuses on linear algebra, calculus, and ordinary differential equations. Additional topics include partial differential equations and approximation methods. Includes solved problems. 1992 edition. 400pp. 7 1/2 x 9 1/4. 0-486-47930-7

PARTIAL DIFFERENTIAL EQUATIONS FOR SCIENTISTS AND ENGINEERS, Stanley J. Farlow. Practical text shows how to formulate and solve partial differential equations. Coverage of diffusion-type problems, hyperbolic-type problems, elliptic-type problems, numerical and approximate methods. Solution guide available upon request. 1982 edition. 414pp. 6 1/8 x 9 1/4. 0-486-67620-X

VARIATIONAL PRINCIPLES AND FREE-BOUNDARY PROBLEMS, Avner Friedman. Advanced graduate-level text examines variational methods in partial differential equations and illustrates their applications to free-boundary problems. Features detailed statements of standard theory of elliptic and parabolic operators. 1982 edition. 720pp. 6 1/8 x 9 1/4. 0-486-47853-X

LINEAR ANALYSIS AND REPRESENTATION THEORY, Steven A. Gaal. Unified treatment covers topics from the theory of operators and operator algebras on Hilbert spaces; integration and representation theory for topological groups; and the theory of Lie algebras, Lie groups, and transform groups. 1973 edition. 704pp. 6 1/8 x 9 1/4. 0-486-47851-3

Browse over 9,000 books at www.doverpublications.com

Mathematics–Logic and Problem Solving

PERPLEXING PUZZLES AND TANTALIZING TEASERS, Martin Gardner. Ninety-three riddles, mazes, illusions, tricky questions, word and picture puzzles, and other challenges offer hours of entertainment for youngsters. Filled with rib-tickling drawings. Solutions. 224pp. 5 3/8 x 8 1/2. 0-486-25637-5

MY BEST MATHEMATICAL AND LOGIC PUZZLES, Martin Gardner. The noted expert selects 70 of his favorite "short" puzzles. Includes The Returning Explorer, The Mutilated Chessboard, Scrambled Box Tops, and dozens more. Complete solutions included. 96pp. 5 3/8 x 8 1/2. 0-486-28152-3

THE LADY OR THE TIGER?: and Other Logic Puzzles, Raymond M. Smullyan. Created by a renowned puzzle master, these whimsically themed challenges involve paradoxes about probability, time, and change; metapuzzles; and self-referentiality. Nineteen chapters advance in difficulty from relatively simple to highly complex. 1982 edition. 240pp. 5 3/8 x 8 1/2. 0-486-47027-X

SATAN, CANTOR AND INFINITY: Mind-Boggling Puzzles, Raymond M. Smullyan. A renowned mathematician tells stories of knights and knaves in an entertaining look at the logical precepts behind infinity, probability, time, and change. Requires a strong background in mathematics. Complete solutions. 288pp. 5 3/8 x 8 1/2. 0-486-47036-9

THE RED BOOK OF MATHEMATICAL PROBLEMS, Kenneth S. Williams and Kenneth Hardy. Handy compilation of 100 practice problems, hints and solutions indispensable for students preparing for the William Lowell Putnam and other mathematical competitions. Preface to the First Edition. Sources. 1988 edition. 192pp. 5 3/8 x 8 1/2. 0-486-69415-1

KING ARTHUR IN SEARCH OF HIS DOG AND OTHER CURIOUS PUZZLES, Raymond M. Smullyan. This fanciful, original collection for readers of all ages features arithmetic puzzles, logic problems related to crime detection, and logic and arithmetic puzzles involving King Arthur and his Dogs of the Round Table. 160pp. 5 3/8 x 8 1/2. 0-486-47435-6

UNDECIDABLE THEORIES: Studies in Logic and the Foundation of Mathematics, Alfred Tarski in collaboration with Andrzej Mostowski and Raphael M. Robinson. This well-known book by the famed logician consists of three treatises: "A General Method in Proofs of Undecidability," "Undecidability and Essential Undecidability in Mathematics," and "Undecidability of the Elementary Theory of Groups." 1953 edition. 112pp. 5 3/8 x 8 1/2. 0-486-47703-7

LOGIC FOR MATHEMATICIANS, J. Barkley Rosser. Examination of essential topics and theorems assumes no background in logic. "Undoubtedly a major addition to the literature of mathematical logic." – *Bulletin of the American Mathematical Society.* 1978 edition. 592pp. 6 1/8 x 9 1/4. 0-486-46898-4

INTRODUCTION TO PROOF IN ABSTRACT MATHEMATICS, Andrew Wohlgemuth. This undergraduate text teaches students what constitutes an acceptable proof, and it develops their ability to do proofs of routine problems as well as those requiring creative insights. 1990 edition. 384pp. 6 1/2 x 9 1/4. 0-486-47854-8

FIRST COURSE IN MATHEMATICAL LOGIC, Patrick Suppes and Shirley Hill. Rigorous introduction is simple enough in presentation and context for wide range of students. Symbolizing sentences; logical inference; truth and validity; truth tables; terms, predicates, universal quantifiers; universal specification and laws of identity; more. 288pp. 5 3/8 x 8 1/2. 0-486-42259-3

Mathematics–Algebra and Calculus

Mathematics–Probability and Statistics

BASIC PROBABILITY THEORY, Robert B. Ash. This text emphasizes the probabilistic way of thinking, rather than measure-theoretic concepts. Geared toward advanced undergraduates and graduate students, it features solutions to some of the problems. 1970 edition. 352pp. 5 3/8 x 8 1/2. 0-486-46628-0

PRINCIPLES OF STATISTICS, M. G. Bulmer. Concise description of classical statistics, from basic dice probabilities to modern regression analysis. Equal stress on theory and applications. Moderate difficulty; only basic calculus required. Includes problems with answers. 252pp. 5 5/8 x 8 1/4. 0-486-63760-3

OUTLINE OF BASIC STATISTICS: Dictionary and Formulas, John E. Freund and Frank J. Williams. Handy guide includes a 70-page outline of essential statistical formulas covering grouped and ungrouped data, finite populations, probability, and more, plus over 1,000 clear, concise definitions of statistical terms. 1966 edition. 208pp. 5 3/8 x 8 1/2. 0-486-47769-X

GOOD THINKING: The Foundations of Probability and Its Applications, Irving J. Good. This in-depth treatment of probability theory by a famous British statistician explores Keynesian principles and surveys such topics as Bayesian rationality, corroboration, hypothesis testing, and mathematical tools for induction and simplicity. 1983 edition. 352pp. 5 3/8 x 8 1/2. 0-486-47438-0

INTRODUCTION TO PROBABILITY THEORY WITH CONTEMPORARY APPLICATIONS, Lester L. Helms. Extensive discussions and clear examples, written in plain language, expose students to the rules and methods of probability. Exercises foster problem-solving skills, and all problems feature step-by-step solutions. 1997 edition. 368pp. 6 1/2 x 9 1/4. 0-486-47418-6

CHANCE, LUCK, AND STATISTICS, Horace C. Levinson. In simple, non-technical language, this volume explores the fundamentals governing chance and applies them to sports, government, and business. "Clear and lively ... remarkably accurate." – *Scientific Monthly*. 384pp. 5 3/8 x 8 1/2. 0-486-41997-5

FIFTY CHALLENGING PROBLEMS IN PROBABILITY WITH SOLUTIONS, Frederick Mosteller. Remarkable puzzlers, graded in difficulty, illustrate elementary and advanced aspects of probability. These problems were selected for originality, general interest, or because they demonstrate valuable techniques. Also includes detailed solutions. 88pp. 5 3/8 x 8 1/2. 0-486-65355-2

EXPERIMENTAL STATISTICS, Mary Gibbons Natrella. A handbook for those seeking engineering information and quantitative data for designing, developing, constructing, and testing equipment. Covers the planning of experiments, the analyzing of extreme-value data; and more. 1966 edition. Index. Includes 52 figures and 76 tables. 560pp. 8 3/8 x 11. 0-486-43937-2

STOCHASTIC MODELING: Analysis and Simulation, Barry L. Nelson. Coherent introduction to techniques also offers a guide to the mathematical, numerical, and simulation tools of systems analysis. Includes formulation of models, analysis, and interpretation of results. 1995 edition. 336pp. 6 1/8 x 9 1/4. 0-486-47770-3

INTRODUCTION TO BIOSTATISTICS: Second Edition, Robert R. Sokal and F. James Rohlf. Suitable for undergraduates with a minimal background in mathematics, this introduction ranges from descriptive statistics to fundamental distributions and the testing of hypotheses. Includes numerous worked-out problems and examples. 1987 edition. 384pp. 6 1/8 x 9 1/4. 0-486-46961-1

Mathematics-Geometry and Topology

PROBLEMS AND SOLUTIONS IN EUCLIDEAN GEOMETRY, M. N. Aref and William Wernick. Based on classical principles, this book is intended for a second course in Euclidean geometry and can be used as a refresher. More than 200 problems include hints and solutions. 1968 edition. 272pp. 5 3/8 x 8 1/2. 0-486-47720-7

TOPOLOGY OF 3-MANIFOLDS AND RELATED TOPICS, Edited by M. K. Fort, Jr. With a New Introduction by Daniel Silver. Summaries and full reports from a 1961 conference discuss decompositions and subsets of 3-space; n-manifolds; knot theory; the Poincaré conjecture; and periodic maps and isotopies. Familiarity with algebraic topology required. 1962 edition. 272pp. 6 1/8 x 9 1/4. 0-486-47753-3

POINT SET TOPOLOGY, Steven A. Gaal. Suitable for a complete course in topology, this text also functions as a self-contained treatment for independent study. Additional enrichment materials make it equally valuable as a reference. 1964 edition. 336pp. 5 3/8 x 8 1/2. 0-486-47222-1

INVITATION TO GEOMETRY, Z. A. Melzak. Intended for students of many different backgrounds with only a modest knowledge of mathematics, this text features self-contained chapters that can be adapted to several types of geometry courses. 1983 edition. 240pp. 5 3/8 x 8 1/2. 0-486-46626-4

TOPOLOGY AND GEOMETRY FOR PHYSICISTS, Charles Nash and Siddhartha Sen. Written by physicists for physics students, this text assumes no detailed background in topology or geometry. Topics include differential forms, homotopy, homology, cohomology, fiber bundles, connection and covariant derivatives, and Morse theory. 1983 edition. 320pp. 5 3/8 x 8 1/2. 0-486-47852-1

BEYOND GEOMETRY: Classic Papers from Riemann to Einstein, Edited with an Introduction and Notes by Peter Pesic. This is the only English-language collection of these 8 accessible essays. They trace seminal ideas about the foundations of geometry that led to Einstein's general theory of relativity. 224pp. 6 1/8 x 9 1/4. 0-486-45350-2

GEOMETRY FROM EUCLID TO KNOTS, Saul Stahl. This text provides a historical perspective on plane geometry and covers non-neutral Euclidean geometry, circles and regular polygons, projective geometry, symmetries, inversions, informal topology, and more. Includes 1,000 practice problems. Solutions available. 2003 edition. 480pp. 6 1/8 x 9 1/4. 0-486-47459-3

TOPOLOGICAL VECTOR SPACES, DISTRIBUTIONS AND KERNELS, François Trèves. Extending beyond the boundaries of Hilbert and Banach space theory, this text focuses on key aspects of functional analysis, particularly in regard to solving partial differential equations. 1967 edition. 592pp. 5 3/8 x 8 1/2. 0-486-45352-9

INTRODUCTION TO PROJECTIVE GEOMETRY, C. R. Wylie, Jr. This introductory volume offers strong reinforcement for its teachings, with detailed examples and numerous theorems, proofs, and exercises, plus complete answers to all odd-numbered end-of-chapter problems. 1970 edition. 576pp. 6 1/8 x 9 1/4. 0-486-46895-X

FOUNDATIONS OF GEOMETRY, C. R. Wylie, Jr. Geared toward students preparing to teach high school mathematics, this text explores the principles of Euclidean and non-Euclidean geometry and covers both generalities and specifics of the axiomatic method. 1964 edition. 352pp. 6 x 9. 0-486-47214-0

Mathematics-History

THE WORKS OF ARCHIMEDES, Archimedes. Translated by Sir Thomas Heath. Complete works of ancient geometer feature such topics as the famous problems of the ratio of the areas of a cylinder and an inscribed sphere; the properties of conoids, spheroids, and spirals; more. 326pp. 5 3/8 x 8 1/2. 0-486-42084-1

THE HISTORICAL ROOTS OF ELEMENTARY MATHEMATICS, Lucas N. H. Bunt, Phillip S. Jones, and Jack D. Bedient. Exciting, hands-on approach to understanding fundamental underpinnings of modern arithmetic, algebra, geometry and number systems examines their origins in early Egyptian, Babylonian, and Greek sources. 336pp. 5 3/8 x 8 1/2. 0-486-25563-8

THE THIRTEEN BOOKS OF EUCLID'S ELEMENTS, Euclid. Contains complete English text of all 13 books of the Elements plus critical apparatus analyzing each definition, postulate, and proposition in great detail. Covers textual and linguistic matters; mathematical analyses of Euclid's ideas; classical, medieval, Renaissance and modern commentators; refutations, supports, extrapolations, reinterpretations and historical notes. 995 figures. Total of 1,425pp. All books 5 3/8 x 8 1/2.
Vol. I: 443pp. 0-486-60088-2
Vol. II: 464pp. 0-486-60089-0
Vol. III: 546pp. 0-486-60090-4

A HISTORY OF GREEK MATHEMATICS, Sir Thomas Heath. This authoritative two-volume set that covers the essentials of mathematics and features every landmark innovation and every important figure, including Euclid, Apollonius, and others. 5 3/8 x 8 1/2. Vol. I: 461pp. 0-486-24073-8
Vol. II: 597pp. 0-486-24074-6

A MANUAL OF GREEK MATHEMATICS, Sir Thomas L. Heath. This concise but thorough history encompasses the enduring contributions of the ancient Greek mathematicians whose works form the basis of most modern mathematics. Discusses Pythagorean arithmetic, Plato, Euclid, more. 1931 edition. 576pp. 5 3/8 x 8 1/2.
0-486-43231-9

CHINESE MATHEMATICS IN THE THIRTEENTH CENTURY, Ulrich Libbrecht. An exploration of the 13th-century mathematician Ch'in, this fascinating book combines what is known of the mathematician's life with a history of his only extant work, the Shu-shu chiu-chang. 1973 edition. 592pp. 5 3/8 x 8 1/2.
0-486-44619-0

PHILOSOPHY OF MATHEMATICS AND DEDUCTIVE STRUCTURE IN EUCLID'S ELEMENTS, Ian Mueller. This text provides an understanding of the classical Greek conception of mathematics as expressed in Euclid's Elements. It focuses on philosophical, foundational, and logical questions and features helpful appendixes. 400pp. 6 1/2 x 9 1/4. 0-486-45300-6

BEYOND GEOMETRY: Classic Papers from Riemann to Einstein, Edited with an Introduction and Notes by Peter Pesic. This is the only English-language collection of these 8 accessible essays. They trace seminal ideas about the foundations of geometry that led to Einstein's general theory of relativity. 224pp. 6 1/8 x 9 1/4. 0-486-45350-2

HISTORY OF MATHEMATICS, David E. Smith. Two-volume history – from Egyptian papyri and medieval maps to modern graphs and diagrams. Non-technical chronological survey with thousands of biographical notes, critical evaluations, and contemporary opinions on over 1,100 mathematicians. 5 3/8 x 8 1/2.
Vol. I: 618pp. 0-486-20429-4
Vol. II: 736pp. 0-486-20430-8

Physics

THEORETICAL NUCLEAR PHYSICS, John M. Blatt and Victor F. Weisskopf. An uncommonly clear and cogent investigation and correlation of key aspects of theoretical nuclear physics by leading experts: the nucleus, nuclear forces, nuclear spectroscopy, two-, three- and four-body problems, nuclear reactions, beta-decay and nuclear shell structure. 896pp. 5 3/8 x 8 1/2. 0-486-66827-4

QUANTUM THEORY, David Bohm. This advanced undergraduate-level text presents the quantum theory in terms of qualitative and imaginative concepts, followed by specific applications worked out in mathematical detail. 655pp. 5 3/8 x 8 1/2. 0-486-65969-0

ATOMIC PHYSICS AND HUMAN KNOWLEDGE, Niels Bohr. Articles and speeches by the Nobel Prize–winning physicist, dating from 1934 to 1958, offer philosophical explorations of the relevance of atomic physics to many areas of human endeavor. 1961 edition. 112pp. 5 3/8 x 8 1/2. 0-486-47928-5

COSMOLOGY, Hermann Bondi. A co-developer of the steady-state theory explores his conception of the expanding universe. This historic book was among the first to present cosmology as a separate branch of physics. 1961 edition. 192pp. 5 3/8 x 8 1/2. 0-486-47483-6

LECTURES ON QUANTUM MECHANICS, Paul A. M. Dirac. Four concise, brilliant lectures on mathematical methods in quantum mechanics from Nobel Prize-winning quantum pioneer build on idea of visualizing quantum theory through the use of classical mechanics. 96pp. 5 3/8 x 8 1/2. 0-486-41713-1

THE PRINCIPLE OF RELATIVITY, Albert Einstein and Frances A. Davis. Eleven papers that forged the general and special theories of relativity include seven papers by Einstein, two by Lorentz, and one each by Minkowski and Weyl. 1923 edition. 240pp. 5 3/8 x 8 1/2. 0-486-60081-5

PHYSICS OF WAVES, William C. Elmore and Mark A. Heald. Ideal as a classroom text or for individual study, this unique one-volume overview of classical wave theory covers wave phenomena of acoustics, optics, electromagnetic radiations, and more. 477pp. 5 3/8 x 8 1/2. 0-486-64926-1

THERMODYNAMICS, Enrico Fermi. In this classic of modern science, the Nobel Laureate presents a clear treatment of systems, the First and Second Laws of Thermodynamics, entropy, thermodynamic potentials, and much more. Calculus required. 160pp. 5 3/8 x 8 1/2. 0-486-60361-X

QUANTUM THEORY OF MANY-PARTICLE SYSTEMS, Alexander L. Fetter and John Dirk Walecka. Self-contained treatment of nonrelativistic many-particle systems discusses both formalism and applications in terms of ground-state (zero-temperature) formalism, finite-temperature formalism, canonical transformations, and applications to physical systems. 1971 edition. 640pp. 5 3/8 x 8 1/2. 0-486-42827-3

QUANTUM MECHANICS AND PATH INTEGRALS: Emended Edition, Richard P. Feynman and Albert R. Hibbs. Emended by Daniel F. Styer. The Nobel Prize–winning physicist presents unique insights into his theory and its applications. Feynman starts with fundamentals and advances to the perturbation method, quantum electrodynamics, and statistical mechanics. 1965 edition, emended in 2005. 384pp. 6 1/8 x 9 1/4. 0-486-47722-3

Physics

INTRODUCTION TO MODERN OPTICS, Grant R. Fowles. A complete basic undergraduate course in modern optics for students in physics, technology, and engineering. The first half deals with classical physical optics; the second, quantum nature of light. Solutions. 336pp. 5 3/8 x 8 1/2. 0-486-65957-7

THE QUANTUM THEORY OF RADIATION: Third Edition, W. Heitler. The first comprehensive treatment of quantum physics in any language, this classic introduction to basic theory remains highly recommended and widely used, both as a text and as a reference. 1954 edition. 464pp. 5 3/8 x 8 1/2. 0-486-64558-4

QUANTUM FIELD THEORY, Claude Itzykson and Jean-Bernard Zuber. This comprehensive text begins with the standard quantization of electrodynamics and perturbative renormalization, advancing to functional methods, relativistic bound states, broken symmetries, nonabelian gauge fields, and asymptotic behavior. 1980 edition. 752pp. 6 1/2 x 9 1/4. 0-486-44568-2

FOUNDATIONS OF POTENTIAL THERY, Oliver D. Kellogg. Introduction to fundamentals of potential functions covers the force of gravity, fields of force, potentials, harmonic functions, electric images and Green's function, sequences of harmonic functions, fundamental existence theorems, and much more. 400pp. 5 3/8 x 8 1/2. 0-486-60144-7

FUNDAMENTALS OF MATHEMATICAL PHYSICS, Edgar A. Kraut. Indispensable for students of modern physics, this text provides the necessary background in mathematics to study the concepts of electromagnetic theory and quantum mechanics. 1967 edition. 480pp. 6 1/2 x 9 1/4. 0-486-45809-1

GEOMETRY AND LIGHT: The Science of Invisibility, Ulf Leonhardt and Thomas Philbin. Suitable for advanced undergraduate and graduate students of engineering, physics, and mathematics and scientific researchers of all types, this is the first authoritative text on invisibility and the science behind it. More than 100 full-color illustrations, plus exercises with solutions. 2010 edition. 288pp. 7 x 9 1/4. 0-486-47693-6

QUANTUM MECHANICS: New Approaches to Selected Topics, Harry J. Lipkin. Acclaimed as "excellent" (*Nature*) and "very original and refreshing" (*Physics Today*), these studies examine the Mössbauer effect, many-body quantum mechanics, scattering theory, Feynman diagrams, and relativistic quantum mechanics. 1973 edition. 480pp. 5 3/8 x 8 1/2. 0-486-45893-8

THEORY OF HEAT, James Clerk Maxwell. This classic sets forth the fundamentals of thermodynamics and kinetic theory simply enough to be understood by beginners, yet with enough subtlety to appeal to more advanced readers, too. 352pp. 5 3/8 x 8 1/2. 0-486-41735-2

QUANTUM MECHANICS, Albert Messiah. Subjects include formalism and its interpretation, analysis of simple systems, symmetries and invariance, methods of approximation, elements of relativistic quantum mechanics, much more. "Strongly recommended." – *American Journal of Physics.* 1152pp. 5 3/8 x 8 1/2. 0-486-40924-4

RELATIVISTIC QUANTUM FIELDS, Charles Nash. This graduate-level text contains techniques for performing calculations in quantum field theory. It focuses chiefly on the dimensional method and the renormalization group methods. Additional topics include functional integration and differentiation. 1978 edition. 240pp. 5 3/8 x 8 1/2. 0-486-47752-5

Physics

MATHEMATICAL TOOLS FOR PHYSICS, James Nearing. Encouraging students' development of intuition, this original work begins with a review of basic mathematics and advances to infinite series, complex algebra, differential equations, Fourier series, and more. 2010 edition. 496pp. 6 1/8 x 9 1/4. 0-486-48212-X

TREATISE ON THERMODYNAMICS, Max Planck. Great classic, still one of the best introductions to thermodynamics. Fundamentals, first and second principles of thermodynamics, applications to special states of equilibrium, more. Numerous worked examples. 1917 edition. 297pp. 5 3/8 x 8. 0-486-66371-X

AN INTRODUCTION TO RELATIVISTIC QUANTUM FIELD THEORY, Silvan S. Schweber. Complete, systematic, and self-contained, this text introduces modern quantum field theory. "Combines thorough knowledge with a high degree of didactic ability and a delightful style." – *Mathematical Reviews.* 1961 edition. 928pp. 5 3/8 x 8 1/2. 0-486-44228-4

THE ELECTROMAGNETIC FIELD, Albert Shadowitz. Comprehensive undergraduate text covers basics of electric and magnetic fields, building up to electromagnetic theory. Related topics include relativity theory. Over 900 problems, some with solutions. 1975 edition. 768pp. 5 5/8 x 8 1/4. 0-486-65660-8

THE PRINCIPLES OF STATISTICAL MECHANICS, Richard C. Tolman. Definitive treatise offers a concise exposition of classical statistical mechanics and a thorough elucidation of quantum statistical mechanics, plus applications of statistical mechanics to thermodynamic behavior. 1930 edition. 704pp. 5 5/8 x 8 1/4. 0-486-63896-0

INTRODUCTION TO THE PHYSICS OF FLUIDS AND SOLIDS, James S. Trefil. This interesting, informative survey by a well-known science author ranges from classical physics and geophysical topics, from the rings of Saturn and the rotation of the galaxy to underground nuclear tests. 1975 edition. 320pp. 5 3/8 x 8 1/2. 0-486-47437-2

STATISTICAL PHYSICS, Gregory H. Wannier. Classic text combines thermodynamics, statistical mechanics, and kinetic theory in one unified presentation. Topics include equilibrium statistics of special systems, kinetic theory, transport coefficients, and fluctuations. Problems with solutions. 1966 edition. 532pp. 5 3/8 x 8 1/2. 0-486-65401-X

SPACE, TIME, MATTER, Hermann Weyl. Excellent introduction probes deeply into Euclidean space, Riemann's space, Einstein's general relativity, gravitational waves and energy, and laws of conservation. "A classic of physics." – *British Journal for Philosophy and Science.* 330pp. 5 3/8 x 8 1/2. 0-486-60267-2

RANDOM VIBRATIONS: Theory and Practice, Paul H. Wirsching, Thomas L. Paez and Keith Ortiz. Comprehensive text and reference covers topics in probability, statistics, and random processes, plus methods for analyzing and controlling random vibrations. Suitable for graduate students and mechanical, structural, and aerospace engineers. 1995 edition. 464pp. 5 3/8 x 8 1/2. 0-486-45015-5

PHYSICS OF SHOCK WAVES AND HIGH-TEMPERATURE HYDRO DYNAMIC PHENOMENA, Ya B. Zel'dovich and Yu P. Raizer. Physical, chemical processes in gases at high temperatures are focus of outstanding text, which combines material from gas dynamics, shock-wave theory, thermodynamics and statistical physics, other fields. 284 illustrations. 1966–1967 edition. 944pp. 6 1/8 x 9 1/4. 0-486-42002-7